State Machines using VHDL

Orhan Gazi • A. Çağrı Arlı

State Machines using VHDL

FPGA Implementation of Serial
Communication and Display Protocols

 Springer

Orhan Gazi
Çankaya University
Ankara, Turkey

A. Çağrı Arlı
Electra IC
Ankara, Turkey

ISBN 978-3-030-61700-4   ISBN 978-3-030-61698-4   (eBook)
https://doi.org/10.1007/978-3-030-61698-4

This Springer imprint is published by the registered company Springer Nature Switzerland AG
The registered company address is: Gewerbestrasse 11, 6330 Cham, Switzerland

Preface

The invention of state machines enabled the development of digital electronic devices. State machines can be considered as mathematical modeling of systems, and these systems can vary from biological organisms, mechanical and electronic devices to human societies. Systems are organized units and get information from environment and process or interpret the information and usually produce a response. Even the universe can be thought as a large state machine made up of state machines large and small.

Electronic engineers are interested in systems designed using digital electronic components. With the evolution of technology in time, the complexity of the systems increased tremendously. Manual design of digital electronic devices used to be performed in past; however, in today's technology, it is almost impossible to design complex systems manually. Even computer programs used for graphical design are useless. Recently hardware design languages such as VHDL and Verilog gained popularity in electronic world. Using hardware design languages, it is possible to design digital electronic devices even consisting of millions of digital gates. In today's technology, the trend is to the use of reconfigurable high-speed digital electronic devices. At the time this book was being written, 5G communication technology was to be adopted by many countries, and for 5G standard many of the digital systems were designed using VHDL hardware design language and FPGAs devices. It is not exaggerating to say that the FPGAs will be the most dominant components of the future digital electronic devices, and especially they will be vital parts of high-speed communication devices. State machines are very widely used to implement algorithms or circuits in VHDL. A digital design engineer should have strong knowledge of state machines and its implementation using hardware design languages. This can even be considered as a vital skill for hardware design engineers.

In this book, we provide information about state machines and VHDL programming using state machines. In Chap. 1, state machine concept is explained, and variety of examples are provided for the mathematical modeling of systems using state machines. We suggest the reader to study the Chap. 1 before proceeding to the other chapters. In Chap. 2, first, we deliver and explain the templates used for the implementation of state machines in VHDL, and then provide example implementations. The examples written for the VHDL implementation of state machines in Chap. 2 contain both theoretical problems and practical applications. We also provided test benches for some of the examples to simulate the written programs in hardware language development platforms. Asynchronous serial communication implementation is given as an example for the implementation of practical applications.

VHDL implementation using timed state machine is explained in Chap. 3. Timed state machines can be considered as a general form of classical state machines. Many of the interfacing design made using VHDL are nothing but applications of timed state machines to practical problems. For this reason, we advise the reader to comprehend the implementation of timed state machines explained in Chap. 3. Once the reader has full comprehension of the timed state machines and their implementations, it is quite straight to perform its applications for interface designs using VHDL.

In Chaps. 4 and 5, the implementation of synchronous serial communication protocols, serial peripheral interface (SPI), and inter-integrated circuit (I2C) communication, in VHDL are explained. The examples provided in this chapter can be considered as the use of timed state machines for the implementation of serial communication protocols developed by some electronic companies for device to devices short distance communication. The readers can use the VHDL codes of this chapter with some modifications for their works. At the end of the Chaps. 4 and 5, we provided practical examples for the implementation of SPI and I2C protocols.

In Chap. 6, different from Chaps. 4 and 5, we consider the implementation of video graphics array (VGA) display protocol developed for computer to monitor image/video transmission. We did not use the state machines for the implementation of VGA display protocol, although they can be used. In Chap. 6, we also considered the data transmission using a high definition multimedia interface (HDMI) which is used for the transmission of video and audio data. The components of HDMI are implemented in VHDL and used for the transmission of image data and VGA display control signals.

The subjects provided in this book can be studied by anyone independently or they can be taught in one semester's course. This book is not a preliminary VHDL tutorial book. We assume that the reader has a basic knowledge of VHDL programming and FPGAs.

Ankara, Turkiye   Orhan Gazi
Ankara, Turkiye   A. Çağrı Arlı

Acknowledgments

There is no doubt to say that state machines are the souls of digital electronic devices. I delivered courses related to VHDL circuit design and state machines for years. The writing of this book became possible after collecting lecture notes and examples from the delivered courses over many years. I would like to thank those students who showed great will in participating in my lectures, their interests motivated me for the writing of this book. I would like to thank Dr. Ahmet Çağrı Arlı who provided and tested many of the VHDL examples and wrote some parts of this textbook. I am dedicating this book to my lovely daughter Vera Gazi who was 7+ years old when this book is being written. Her love was a motivating factor for all my studies.

Dr. Orhan Gazi

I hope this book will be useful to prospective engineers and engineers around the world. I have no doubt that the subject of the book is up-to-date and will teach readers basic interface protocols. As the greatest effort in creating the book, I would like to thank my esteemed teacher, Orhan Gazi, for the way and efforts he made in writing the book. I thank my father Fahrettin, who is also a textbook author, for the inspiration he gave to me. In addition, I am grateful to my grandfather Ahmet, who is also an author, for being an example to me with his disciplined work with his successful creation of our family memories. I would like to thank my mother, Devrim, who, I know, was always there for me. Finally, I am grateful to my love, Görkem, for her supports during the writing of this book.

Dr. Ahmet Çağrı Arlı

Abbreviations

ACK	Acknowledgement
ADC	Analog to digital converter
CEC	Consumer electronics control
CLK	Clock
CPHA	Clock phase
CPOL	Clock polarity
CRT	Cathode ray tubes
DAC	Digital to analog converter
DCL	Data clock
DDC	Display data channel
DVI	Digital visual interface
FPGA	Field-programmable gate array
HACTIVE	Horizontal active
HDMI	High-definition multimedia interface
HSYNC	Horizontal synchronization
I2C	Inter integrated circuit
LED	Light emitting diode
MISO	Master-input slave-output
MOSI	Master-output slave-input
NS	Next state
PS	Present state
SCLK	Serial clock
SDA	Serial data
SI	Serial input
SO	Serial output
SPI	Serial peripheral interface
SS	Slave select
TB	Test-bench
TMDS	Transition-minimized differential signaling
VACTIVE	Vertical active
VGA	Video graphics array
VHDL	Very high speed integrated circuit hardware description language
VSYNC	Vertical synchronization

Contents

State Machines and Modeling of Mathematical and Physical Problems by State Machines

1

State machines are used to characterize the behavior of a system which can be an electronic circuit, or a real-life event, and a physical event. Even the psychology of a human can be modeled by a state machine. The invention of state machines was a reason for the development of computers. State machines are widely used in industry. For instance, they are used in factories for control applications. Communication technology heavily relies on the state machines. Without the availability of state machines, the existence of today's technology would not be possible. In this chapter, we provide information about the state machines, and state machine modeling of physical and abstract events.

1.1   State

The behavior of a circuit involving memory units can be expressed using states. The outputs of the memory elements can be considered as states. On the other hand, states are also defined as systems used in real-life. The state change occurs if an input is applied to the system. The human body can also be considered as a system. When you receive a good news, your mode changes to happiness which can be considered as a state, on the other hand, when you receive a bad news, your mode changes to sorry which can be considered as another state.

The states used for digital electronic circuits are changed in a controlled manner. The controller is the clock course. When an input is applied, state change does not occur immediately, but a clock pulse application waits. Upon the application of a clock pulse, state change occurs.

The changes between states can be illustrated using flow charts, state tables, or state diagrams. The most common representation is the state diagram.

O. Gazi, A. Ç. Arlı, *State Machines using VHDL*,
https://doi.org/10.1007/978-3-030-61698-4_1

1.1.1   State Diagrams and Mealy and Moore Models

State diagrams are used to express the behavior of sequential circuits graphically. State diagrams contain the same information as state tables. The information available in a state table can be expressed using state diagrams. Transitions between states are shown considering the inputs and outputs of the circuit as depicted in Fig. 1.1. Transitions between states occur upon the application of a clock pulse.

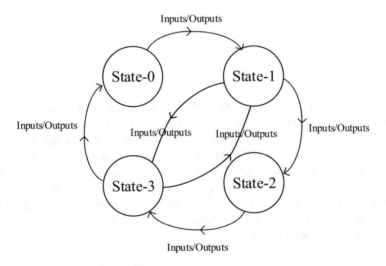

Fig. 1.1 A typical Mealy state diagram

1.1.1.1 Mealy and Moore Machines

Finite state machines can be divided into two main categories which are named as Mealy state machines and Moore state machines.

The outputs of the Moore state machines depend solely only on the present state values, on the other hand, the outputs of the Mealy machines depend on both present states and external inputs. The general structure of Moore state machines is depicted in Fig. 1.2.

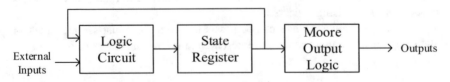

Fig. 1.2 The general structure of Moore state machines

The general structure of Mealy state machines is depicted in Fig. 1.3.

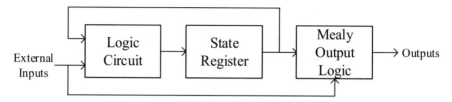

Fig. 1.3 The general structure of Mealy state machines

We can implement a sequential circuit using Mealy or Moore machines, and a state diagram for a Mealy machine can be converted to a state diagram of Moore machine and vice versa. If we want to explicitly emphasize the type of the state machine, then for Moore machines the outputs of the circuits are written under the name of the state identity as in Fig. 1.4.

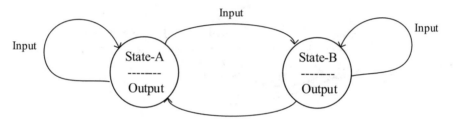

Fig. 1.4 Generic model for Moore state machine

The general state diagram for a Mealy state machine is depicted in Fig. 1.5.

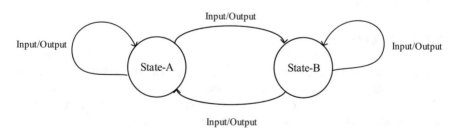

Fig. 1.5 Generic Model for Mealy state machine

Example 1.1 In Fig. 1.6, state diagram of a Mealy state machine is depicted.

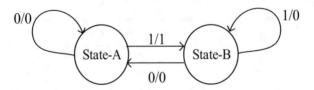

Fig. 1.6 Mealy state diagram for Example 1.1

The Mealy state diagram of Fig. 1.6 can be converted to a Moore state diagram as in Fig. 1.7.

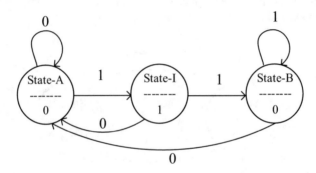

Fig. 1.7 Moore conversion result

Note that different circuits are synthesized for Moore and Mealy state machines even though they do the same task.

1.1.2   State Names

The names are assigned to the states in Mealy and Moore models. Since, in the Moore models the outputs are indicated below the state names, for the simplicity of illustration, the state names can be omitted, and in this case state outputs are also used for the names of the states. For Mealy models, state names cannot be omitted.

Example 1.2 The Moore state diagram in Fig. 1.8a can also be drawn as in Fig. 1.8b.

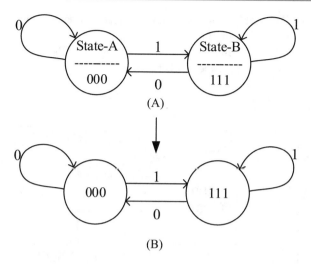

(A)

(B)

Fig. 1.8 Moore state diagram representations

1.1.3   State Machine Inputs and Outputs

The inputs and outputs in a state diagram can also be indicated using letters or words. If a letter word has an apostrophe, it indicates that the input value indicated by letter or word equals to 0, otherwise it is equal to 1.

Example 1.3 Two Moore state diagrams which employ letters and words for inputs or outputs are depicted in Fig. 1.9.

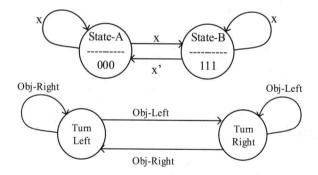

Fig. 1.9 Two Moore state diagrams

1.2   Modeling of Mathematical and Real-Life Problems by State Machines

In this section, we provide examples for the modeling of mathematical and real-life problems by state machines. We will consider both Mealy and Moore state diagrams for modeling.

Example 1.4 The block diagram of an obstacle avoidance robot is shown in Fig. 1.10. When an object on the right is detected, only the right motor works and the robot turns left. In a similar manner, when an object on the left is detected, only the left motor works and the robot turns right.

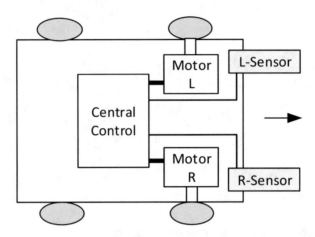

Fig. 1.10 The block diagram of obstacle avoidance robot

The operation of the obstacle avoidance robot can be described using Moore state machine in Fig. 1.11.

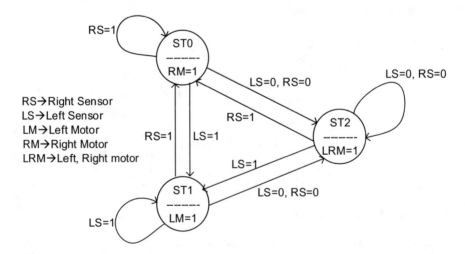

Fig. 1.11 State diagram of obstacle avoidance robot

The state names with their outputs and inputs of the Moore state diagram of Fig. 1.11 can be expressed using words in a meaningful manner as in Fig. 1.12.

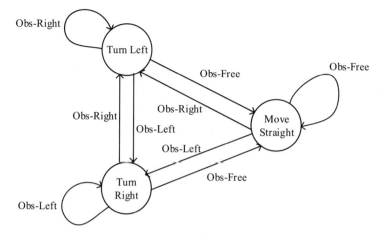

Fig. 1.12 State diagram of obstacle avoidance robot

Example 1.5 Obtain the Moore state diagram of the electronic counter circuit which works as follows. When the external input is 0, the counter follows the count sequence 0–2–4–6. On the other hand, when the external input is 1, the counter follows the count sequence 1–3–5–7–0.

Solution 1.5 Considering the given information, we can construct the Moore state diagram as in Fig. 1.13 where x denotes the external input, and state names are the same as the state outputs, i.e., circuit outputs.

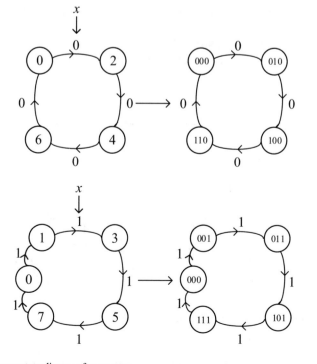

Fig. 1.13 Moore state diagram for counter

Example 1.6 Obtain the state diagram of the circuit which detects three consecutive 1's in a bit string. Overlapping is allowed.

Solution 1.6 The Mealy state diagram of three 1's detector is depicted in Fig. 1.14 where it is indicated that the circuit output equals to 1 when three consecutive 1's are detected otherwise circuit output equals to 0.

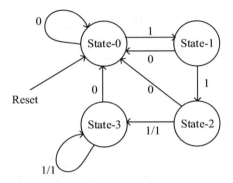

Fig. 1.14 Mealy state diagram for three consecutive 1's detector

For instance, for the input stream 10111 10111 001 the circuit output is 00001 10001 000.

Example 1.7 Obtain the state diagram of non-return to zero inverted (NRZI) encoding technique. In this technique, the circuit inputs a bit string and performs encoding operation. The encoding is done in a way such that, if the input is a 0, then no change is performed on the output signal, otherwise complement of the output is taken.

Solution 1.7 In Fig. 1.15, the operation of NRZI technique is illustrated.

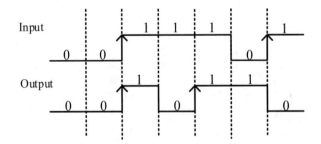

Fig. 1.15 Non-return to zero inverted (NRZI) encoding

The corresponding Mealy state diagram is depicted in Fig. 1.16.

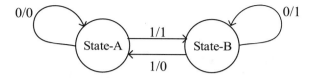

Fig. 1.16 Mealy state diagram for NRZI encoding

Example 1.8 The bit stuffing circuit inserts a 0 whenever three consecutive 1's are detected in a bit string. Obtain the state diagram of the circuit.

Solution 1.8 The circuit output equals 0 whenever three consecutive 1's are detected, otherwise, the circuit output is the same as the input bit. For instance, for the input stream 110 111 1111 0110 the circuit output is 110 111 **0** 111 **0** 1 0110.

The bit stuffing operation is graphically illustrated in Fig. 1.17.

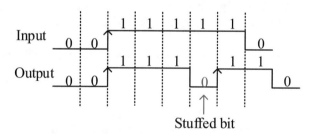

Fig. 1.17 Bit stuffing waveforms

Considering the given information, we can draw the Mealy state diagram of the bit stuffing circuit as in Fig. 1.18.

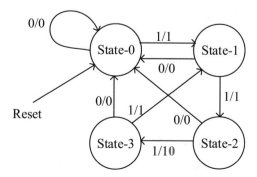

Fig. 1.18 Mealy state diagram of the bit stuffing operation

The Moore state diagram of the bit stuffing circuit is depicted in Fig. 1.19.

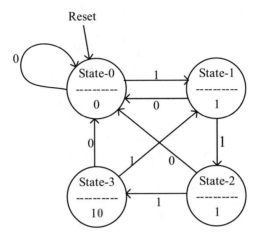

Fig. 1.19 Moore state diagram of the bit stuffing operation

Example 1.9 The rising edge detector generates logic-1 when a transition from 0 to 1 is detected. Obtain the state diagram of the edge detector.

Solution 1.9 The operation of the edge detector is graphically illustrated in Fig. 1.20.

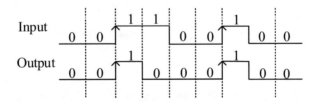

Fig. 1.20 Input/output waveforms of edge detector

Considering the given information, we can draw the Mealy state diagram of the edge detector as in Fig. 1.21.

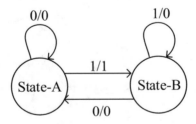

Fig. 1.21 Mealy state diagram for edge detector

Example 1.10 An elevator is used between two floors. The elevator has up and down buttons, and inside the elevator there are one green and one red light. At down floor only green light is on and at up floor only red light is on. Obtain the state diagram for the elevator machine.

Solution 1.10 The inputs of the elevator are the up and down buttons, and the outputs are the green and red lights. Considering the given information, we can draw the state machine of the elevator as in Fig. 1.22 where letters are used to denote the inputs and outputs, and if a letter has an apostrophe, it indicates that 0 is the value of the letter, otherwise, it has value 1.

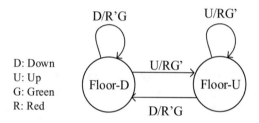

Fig. 1.22 Mealy state diagram of the elevator machine

Example 1.11 Obtain the state diagram of the Manchester encoding illustrated in Fig. 1.23.

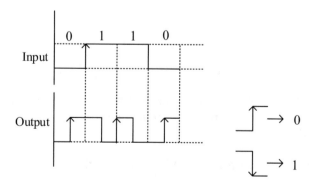

Fig. 1.23 Manchester encoding waveforms

Solution 1.11 Pulse levels can be expressed using two binary digits as shown in Fig. 1.24.

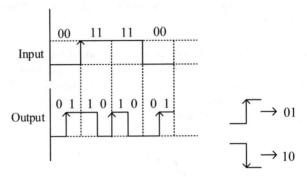

Fig. 1.24 Manchester encoding

Considering Fig. 1.24, the state diagram of the Manchester encoding can be drawn as in Fig. 1.25a which can also be drawn as in Fig. 1.25b.

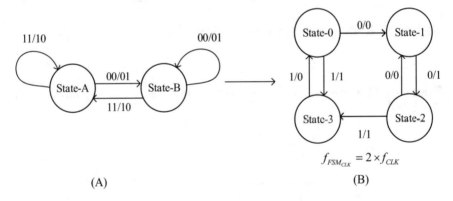

(A)   (B)

Fig. 1.25 State diagrams for Manchester encoding

Since pulse levels are expressed using two bits, the clock frequency, i.e., the clock speed of the state machine, is twice that of the bit clock.

1.2.1   Some Applications of Finite State Machines

Some practical application of state machines can be seen in: Vending Machines, Traffic Lights, Video Games, Memory Controllers, Communication Protocols, Speech Processing, Channel Encoding, Industrial Control, etc.

The following example illustrates the use of state machines in asynchronous serial communication.

Example 1.12 RS232 signaling shown in Fig. 1.26 is used for asynchronous data transmission. Obtain the state diagram of the RS232 transmission waveform.

| | Start | Bit-0 | Bit-1 | Bit-2 | Bit-3 | Bit-4 | Bit-5 | Bit-6 | Bit-7 | Stop | |

Fig. 1.26 RS232 transmission waveform

Solution 1.12 The transmission waveform shown in Fig. 1.26 can be expressed using Moore state machine as in Fig. 1.27 where transmission of each pulse is considered as a separate state.

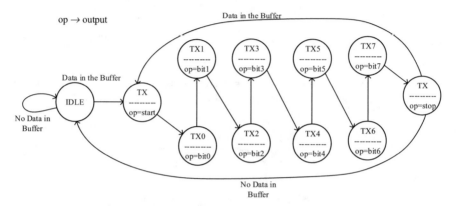

Fig. 1.27 State diagram of RS232 transmission waveform

The Moore state diagram of Fig. 1.27 can be drawn as in Fig. 1.28 where inside states we only use a single word, and each of these words imply the detailed information shown in the states of Fig. 1.27.

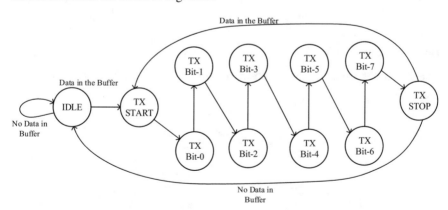

Fig. 1.28 Alternative representation of state diagram of RS232 transmission waveform

Example 1.13 Design a simple sequence detector for the detection of the sequence 011. Include three outputs that indicate how many bits have been received in the correct sequence.

(Each output is connected to a LED.) Draw the state diagram for Moore state machine.

Solution 1.13 We will use Moore state machines for this example. The initial state is shown in Fig. 1.29 where the state output written below the state name indicates whether the sequence is detected or not. When the sequence is detected, state output equals to logic-1.

Fig. 1.29 Starting state

If the first input bit is 0, a transition is made to "State-B", on the other hand, if the first input bit is 1, a transition to "State-A" itself is made as shown in Fig. 1.30.

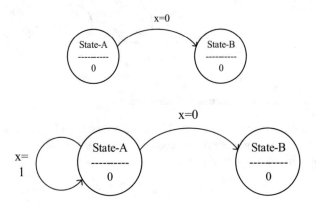

Fig. 1.30 Transitions from "State-A"

The transitions from "State-B" considering different input values are drawn as in Fig. 1.31.

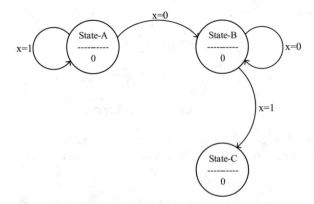

Fig. 1.31 Transitions from "State-B"

The transitions from "State-C" considering different input values are drawn in Fig. 1.32 where it is clear that when "State-D" is reached, the sequence 011 is detected, and output value equals to 1.

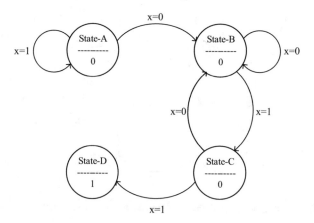

Fig. 1.32 Transitions from 'State-C'

If the input bit is 0 at "State-D", a transition is made to "State-B" as shown in Fig. 1.33.

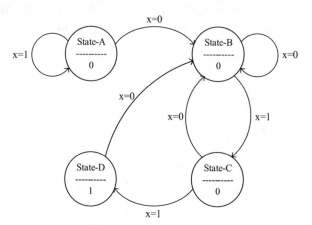

Fig. 1.33 Transitions from "State-D"

If the input bit is 1 at "State-D", a transition is made to the initial state "State-A" as shown in Fig. 1.34 which is the complete state diagram.

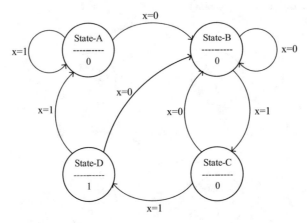

Fig. 1.34 Moore state machine diagram for the detection of 011 sequence

Example 1.14 Design a sequence detector that searches the pattern $01[0*]1$, where $[0*]$ is any number of consecutive zeroes, in a bit stream. The output become equal to 1 every time the pattern is detected.

Draw the Mealy state diagram for this sequence detector.

Solution 1.14 From every state, there are two outgoing transitions corresponding to inputs 0 and 1. Considering this, the transitions from the starting state can be drawn as in Fig. 1.35.

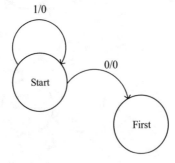

Fig. 1.35 Transitions from state 'Start'

The transitions for the inputs 0 and 1 from the state "First" are drawn as in Fig. 1.36. The state names except for "Delay" indicate the number of bits found correctly for the pattern $01[0*]1$.

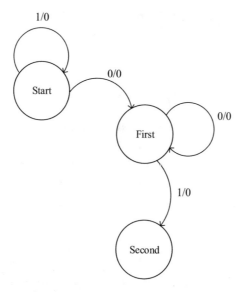

Fig. 1.36 Transitions from state "First"

The transitions for the inputs 0 and 1 from the state "Second" are drawn as in Fig. 1.37.

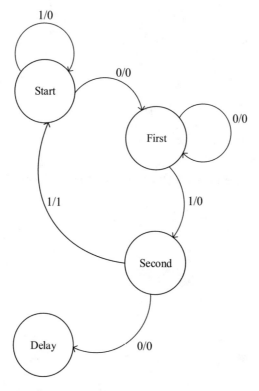

Fig. 1.37 Transitions from state "Second"

The transitions for the inputs 0 and 1 from the state "Delay" are drawn as in Fig. 1.38 where the state "Success" indicates that the pattern is found.

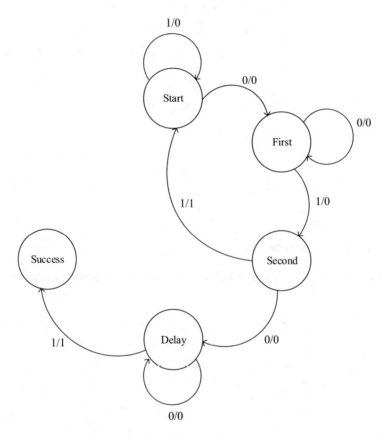

Fig. 1.38 Transitions from state "Delay"

The transitions for the inputs 0 and 1 from the state "Success" are drawn as in Fig. 1.39.

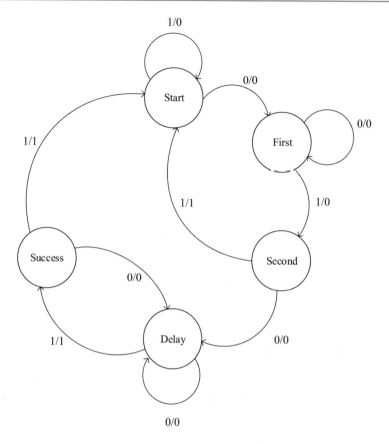

Fig. 1.39 Complete state diagram of sequence detector

Example 1.15 Design a sequence detector that searches the pattern 01[0*]1, where [0*] is any number of consecutive zeroes, in a bit stream. The output become equal to 1 every time the pattern is detected.

Draw the state diagram for Moore machine implementation.

Solution 1.15 The state machines detects the sequence 010[0*]1 where [0*] indicates a sequence of zeros. The shortest sequence to be detected is 011. When the sequence is detected the state output equals to 1. Otherwise, the state output equals to 0. In our Moore model, the state names except for "Delay" indicate the numbers of 1's to be detected in the required bit sequence.

The initial state and the transitions from the initial state for input 0 and 1 are indicated in Fig. 1.40.

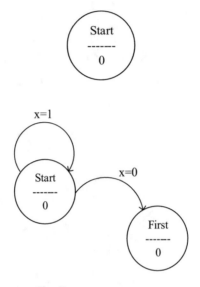

Fig. 1.40 Transitions from state "Start"

The transitions from the state "First" for input 0 and 1 are indicated in Fig. 1.41.

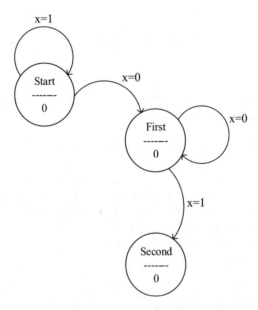

Fig. 1.41 Transitions from state "First"

The transitions from the state "Second" for input 0 and 1 are indicated in Fig. 1.42 where the "Success" state is reached when the input pattern is 011.

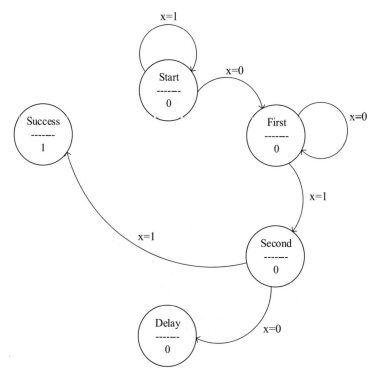

Fig. 1.42 Transitions from state "Second"

The transitions from the state "Delay" for input 0 and 1 are indicated in Fig. 1.43 where an extra "Success-2" is defined, and this state is reached for the pattern 01[0*]1.

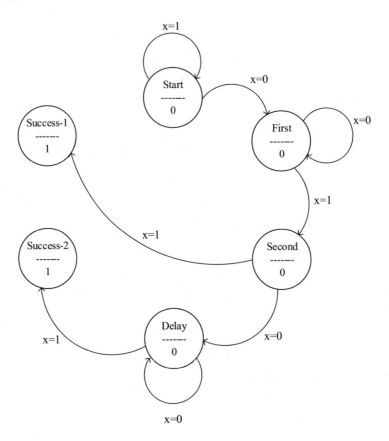

Fig. 1.43 Transitions from state "Delay"

The transitions from the state "Success-2" for input 0 and 1 are indicated in Fig. 1.44.

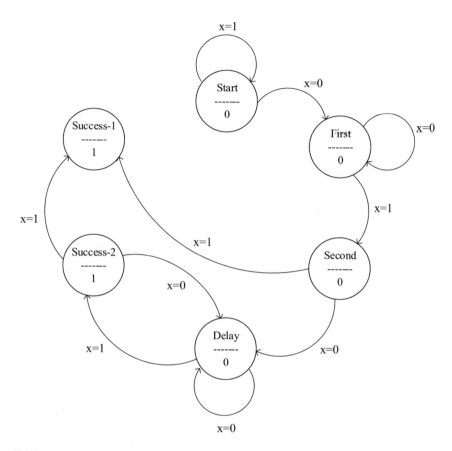

Fig. 1.44 Transitions from state "Success-2"

The transitions from the state "Success-1" for input 0 and 1 are indicated in Fig. 1.45.

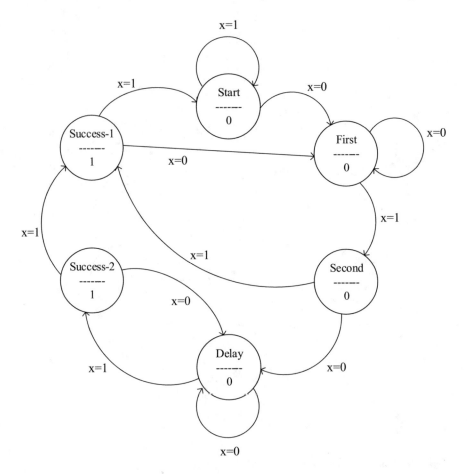

Fig. 1.45 Complete state diagram for Moore model

1.2.2   Mealy or Moore

The Moore machine lags one clock cycle to generate the same output sequence. The Mealy machine can change asynchronously with the input. The Mealy Machine requires one less state than the Moore machine. Mealy machine makes use of more information, i.e., inputs, than Moore machine while forming the outputs. Having a smaller number of states reduces the design cost. In some cases, more than one state reduction is also possible.

However, Mealy machines may face "glitch" problem. Glitches are unwanted temporary outputs appearing for very short time interval, and glitches can be hazardous for electronic devices. To prevent the glitch problems, we need an extra flip-flop in Mealy machines, and in this case, both machines can use the same number of flip-flops.

Example 1.16 Obtain the Moore and Mealy state diagrams for the sequence detector which detects the sequences 0010 or 0001. Overlapping patterns are allowed.

Solution 1.16 Considering the allowance of overlapped patterns, we can draw the Mealy state diagram as in Fig. 1.46.

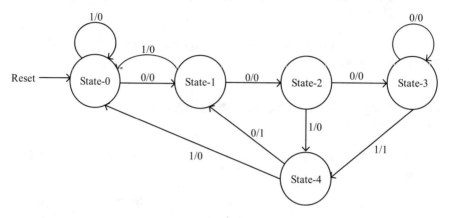

Fig. 1.46 Mealy state diagram for sequence detector

The Moore state diagram can be obtained as in Fig. 1.47. In fact, once you have the Mealy state diagram, you can convert it to Moore state diagram directly.

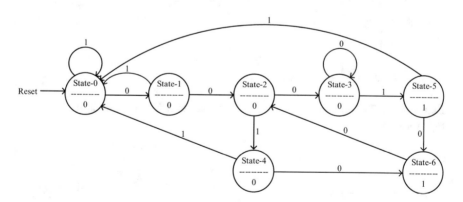

Fig. 1.47 Moore state diagram for sequence detector

For the input sequence $x =$ "10011000010010111010010", the output of the Mealy and Moore machines can be calculated as

$$x:\qquad 1\ 0\ 0\ 1\ 1\ 0\ 0\ 0\ 0\ 1\ 0\ 0\ 1\ 0\ 1\ 1\ 1\ 0\ 1\ 0\ 0\ 1\ 0$$
$$y\ \text{Mealy:}\quad 0\ 0\ 0\ 0\ 0\ 0\ 0\ 0\ 0\ 1\ 1\ 0\ 0\ 1\ 0\ 0\ 0\ 0\ 0\ 0\ 0\ 0\ 1$$
$$y\ \text{Moore:}\quad 0\ 0\ 0\ 0\ 0\ 0\ 0\ 0\ 0\ 0\ 1\ 1\ 0\ 0\ 1\ 0\ 0\ 0\ 0\ 0\ 0\ 0\ 0\ 1$$

where it is seen that Mealy and Moore state diagrams produce the same output sequence for the same input sequence, however, the output of the Moore state diagram is delayed by one clock cycle.

1.3   Conversion Between Mealy and Moore State Diagrams/ Machines

In this section, we will consider the conversion operation between Mealy and Moore state machines/diagrams.

1.3.1   Conversion from Mealy to Moore State Diagrams/ Machines

The conversion of a state in a Mealy model to a state in Moore model is depicted in Fig. 1.48.

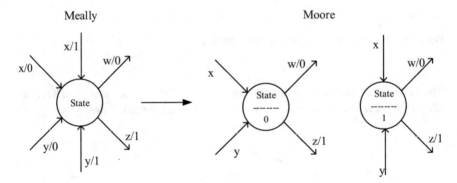

Fig. 1.48 Mealy to Moore conversion operation

Example 1.17 The conversion of a Mealy state to a Moore state is illustrated in Fig. 1.49.

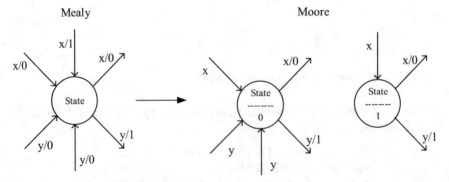

Fig. 1.49 Mealy to Moore conversion example

Example 1.18 Convert the Mealy state machine in Fig. 1.50 to a Moore state machine.

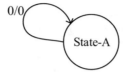

Fig. 1.50 A self-transition state

Solution 1.18 First, let us convert "State-A" as in Fig. 1.51 where it is seen that the output of the initial state is unknown.

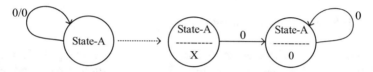

Fig. 1.51 Conversion operation

Example 1.19 Convert the Mealy state machine in Fig. 1.52 to a Moore state machine.

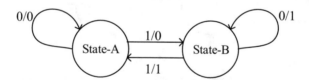

Fig. 1.52 Example Mealy state diagram

Solution 1.19 Assume that starting state is "State-A", i.e., when reset signal is applied to the state machine, starting state happens to be "State-A". First, let us convert "State-A" as in Fig. 1.53.

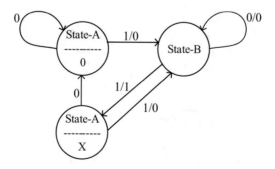

Fig. 1.53 Conversion of "State-A"

The unknown output of "State-A" in Fig. 1.53 can be determined considering the incoming transition from "State-B" as in Fig. 1.54.

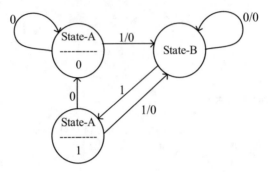

Fig. 1.54 Determination of unknown output

If we convert "State-B", we get the Moore state machine as in Fig. 1.55. Although "State-B" in Fig. 1.55 has self-transitions, it is not the starting state, and for this reason, we do not have two Moore states for "State-B" as in the conversion of "State-A".

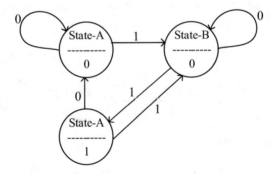

Fig. 1.55 Conversion of "State-B"

Lastly, we can assign different names to each state as in Fig. 1.56.

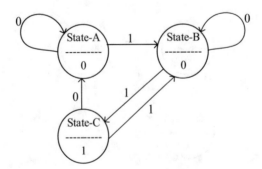

Fig. 1.56 Conversion result

When Mealy and Moore state machines/diagrams are compared, we see that Moore machine has one more state. This means that Moore state machine requires one more flip-flop for its hardware implementation.

Example 1.20 Convert the Mealy state machine shown in Fig. 1.57 to a Moore state machine.

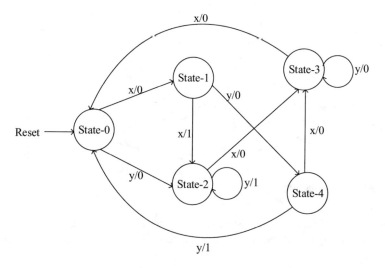

Fig. 1.57 Example Mealy state diagram

Solution 1.20 First, we convert "State-0" for Moore model as in Fig. 1.58.

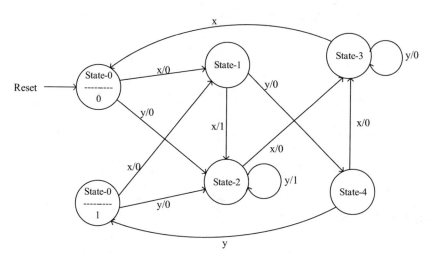

Fig. 1.58 Conversion of "State-0"

In the second step, "State-1" is converted for Moore model as in Fig. 1.59.

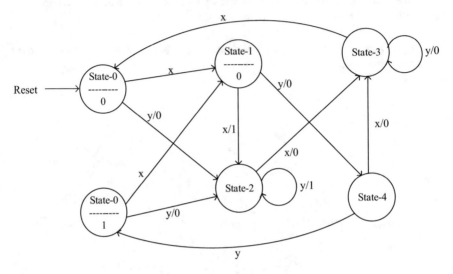

Fig. 1.59 Conversion of "State-1"

Conversions of "State-3" and "State-4" are easier than the conversion of "State-2". For this reason, in the third step, we convert "State-3" for Moore model as in Fig. 1.60.

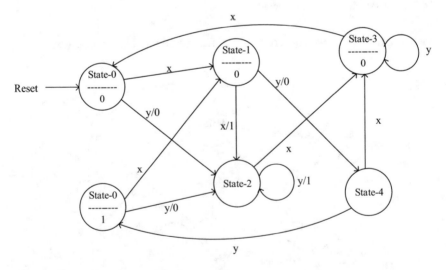

Fig. 1.60 Conversion of "State-3"

In the fourth step, we convert "State-4" for Moore model as in Fig. 1.61.

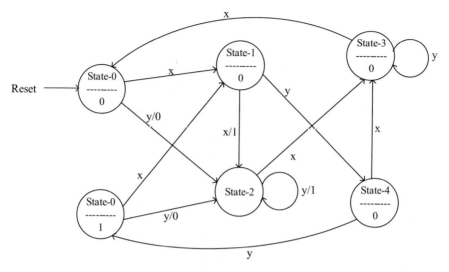

Fig. 1.61 Conversion of "State-4"

Lastly, we convert "State-2" for Moore model as in Fig. 1.62.

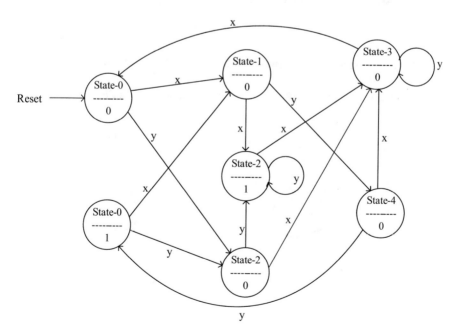

Fig. 1.62 Conversion of "State-2"

We can assign different names to the new states as in Fig. 1.63.

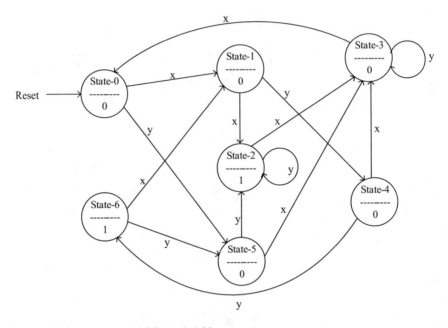

Fig. 1.63 Conversion result of Example 1.20

1.3.2   Conversion from Moore to Mealy State Machines

The conversion of a Moore state to a Mealy state is depicted in Fig. 1.64.

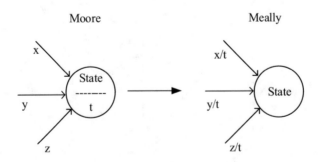

Fig. 1.64 Moore to Mealy conversion operation

Example 1.21 Convert the Moore state machine in Fig. 1.65 to a Mealy state machine.

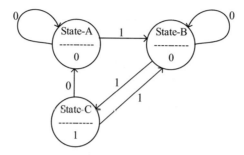

Fig. 1.65 Example state machine for conversion

Solution 1.21 First, we convert "State-A" as in Fig. 1.66.

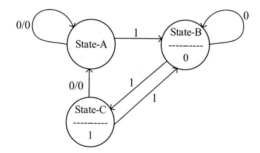

Fig. 1.66 Conversion of "State-A"

The conversion of "State-B" to a Mealy state is depicted in Fig. 1.67.

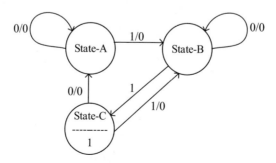

Fig. 1.67 Conversion of "State-B"

The conversion of "State-C" to a Mealy state is depicted in Fig. 1.68.

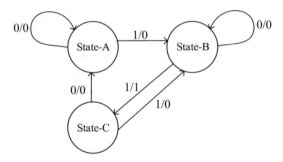

Fig. 1.68 Conversion of "State-C"

When Fig. 1.68 is inspected, we see that for the same inputs, "State-A" and "State-C" produce the same outputs, and these two states are equivalent states and they can be merged. When the equivalent states are merged, we get the final form of the Mealy state machine as in Fig. 1.69.

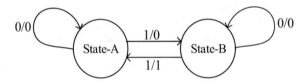

Fig. 1.69 Conversion result after state merging

Example 1.22 Convert the Moore state machine shown in Fig. 1.70 to an equivalent Mealy state machine.

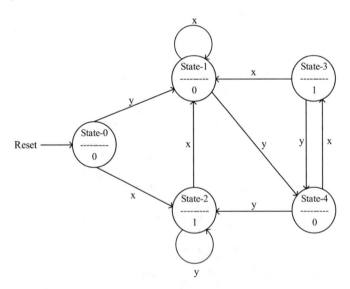

Fig. 1.70 Moore state machine for Example 1.22

Solution 1.22 Following the Moore to Mealy conversion rule depicted in Fig. 1.64, we get the Mealy state diagram shown in Fig. 1.71.

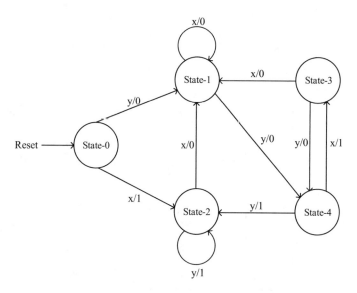

Fig. 1.71 Mealy conversion result for Example 1.22

1.4   Modeling the Behavior of Electronic Circuits Using State Machines

Flip-flops are the memory units, i.e., cells, used for the construction of electronic circuits involving memory elements, and these circuits are controlled by a clock source and their operations are sequential. The outputs of the memory cells are considered as states.

1.4.1   Flip-Flops, Characteristic, and Excitation Tables

The commercially available flip-flops can be listed as D, T, JK, and SR. State tables are used to illustrate the behavior of clocked sequential circuits involving flip-flops. The black box representation of commercially available positive and negative edge triggered D, T, JK, and SR flip-flops are depicted in Figs. 1.72 and 1.73.

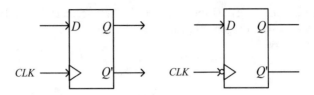

Fig. 1.72 Positive and negative edge triggered D flip-flops

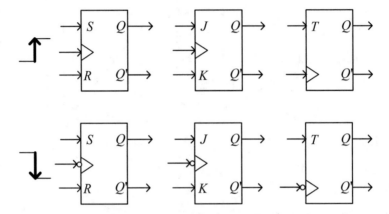

Fig. 1.73 Positive and negative edge triggered SR, JK, and T flip-flops

The behavior of the commercially available flip-flops can be explained using the characteristic and excitation tables of Tables 1.1 and 1.2 where $Q(t)$ is the present state, and $Q(t + 1)$ is the next state, i.e., state after the application clock pulse.

Table 1.1 Characteristic tables of SR, JK, D, and T flip-flops

J	K	$Q(t + 1)$	S	R	$Q(t + 1)$	D	$Q(t + 1)$	T	$Q(t + 1)$
0	0	$Q(t)$	0	0	$Q(t)$	0	0	0	$Q(t)$
0	1	0	0	1	0	1	1	1	$Q'(t)$
1	0	1	1	0	1				
1	1	$Q'(t)$	1	1	X				

Table 1.2 Excitation tables of SR, JK, D, and T flip-flops

$Q(t)$	$Q(t + 1)$	S	R	$Q(t)$	$Q(t + 1)$	J	K
0	0	0	×	0	0	0	×
0	1	1	0	0	1	1	×
1	0	0	1	1	0	×	1
1	1	×	0	1	1	×	0

$Q(t)$	$Q(t + 1)$	D	$Q(t)$	$Q(t + 1)$	T
0	0	0	0	0	0
0	1	1	0	1	1
1	0	0	1	0	1
1	1	1	1	1	0

Example 1.23 The characteristic table of AB flip-flop is given in Table 1.3. Find the excitation table of AB flip-flop.

Table 1.3 The characteristic table of AB flip-flop

A	B	$Q(t+1)$
0	0	$Q'(t)$
0	1	1
1	0	0
1	1	$Q(t)$

Solution 1.23 First we construct the excitation table as in Table 1.4 containing only output values.

Table 1.4 Possible outputs of the flip-flops before and after clock application

$Q(t)$	$Q(t+1)$	A	B
0	0		
0	1		
1	0		
1	1		

For $Q(t) = 0$ and $Q(t+1) = 0$, we can have $=1$, $B = 1$, since $Q(t+1) = Q(t)$ as indicated in Table 1.5.

Table 1.5 The last line of characteristic table of AB flip-flop

A	B	$Q(t+1)$
0	0	$Q'(t)$
0	1	1
1	0	0
1	**1**	$Q(t)$

Besides for $Q(t+1) = 0$ we can have $A = 1$, $B = 0$ as illustrated in Table 1.6.

Table 1.6 The third line of characteristic table of AB flip-flop

A	B	$Q(t+1)$
0	0	$Q'(t)$
0	1	1
1	**0**	0
1	1	$Q(t)$

We conclude that for $Q(t) = 0$ and $Q(t + 1) = 0$ we have $=1$, $B = \times$. Then, the first line of the excitation table happens to be as in Table 1.7.

Table 1.7 The first line of the excitation table of the AB flip-flop

$Q(t)$	$Q(t + 1)$	A	B
0	0	1	×
0	1		
1	0		
1	1		

Proceeding in a similar manner, we obtain the excitation table of the AB flip-flop as in Table 1.8.

Table 1.8 Excitation table of the AB flip-flop

$Q(t)$	$Q(t + 1)$	A	B
0	0	1	×
0	1	0	×
1	0	×	0
1	1	×	1

1.4.2   State Tables and State Diagrams of Sequential Circuits

The header of the state table contains the parameters representing the external inputs, flip-flops' inputs, external outputs, and flip-flops' outputs. The outputs of the flip-flops are decided upon the application of clock pulses considering their input values.

Example 1.24 Obtain the state table of the circuit shown in Fig. 1.74.

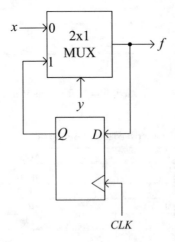

Fig. 1.74 Sequential circuit for Example 1.24

Solution 1.24 In the state table, we first write all the possible combinations of the external inputs and present states as shown in Table 1.9.

Table 1.9 State table construction

x	y	Q(t)	f	Q(t + 1)
0	0	0		
0	0	1		
0	1	0		
0	1	1		
1	0	0		
1	0	1		
1	1	0		
1	1	1		

The multiplexer output can be written as

$$f = y'x + yQ(t)$$

which goes to the input of the D flip-flop, i.e., $D = f$. Upon the application of the clock pulse, the output of the flip-flop is determined as $Q(t + 1) = D$. That means that

$$Q(t+1) = y'x + yQ(t).$$

We can complete the state table as in Table 1.10.

Table 1.10 State table

x	y	Q(t)	f	Q(t + 1)
0	0	0	0	0
0	0	1	0	0
0	1	0	0	0
0	1	1	1	1
1	0	0	1	1
1	0	1	1	1
1	1	0	0	0
1	1	1	1	1

Exercise Obtain the state table of the circuit shown in Fig. 1.75.

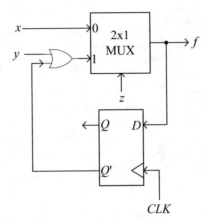

Fig. 1.75 Sequential circuit for exercise

Example 1.25 Obtain the state table of the circuit shown in Fig. 1.76.

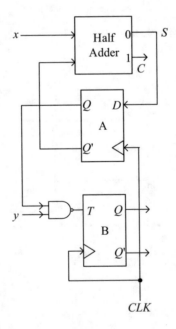

Fig. 1.76 Sequential circuit for Example 1.25

Solution 1.25 First, all the possible values of the external inputs and flip-flop outputs are tabulated as in Table 1.11.

Table 1.11 All possible values of external inputs and flip-flop outputs

x	y	$Q_a(t)$	$Q_b(t)$	D	T	$Q_a(t+1)$	$Q_b(t+1)$
0	0	0	0				
0	0	0	1				
0	0	1	0				
0	0	1	1				
0	1	0	0				
0	1	0	1				
0	1	1	0				
0	1	1	1				
1	0	0	0				
1	0	0	1				
1	0	1	0				
1	0	1	1				
1	1	0	0				
1	1	0	1				
1	1	1	0				
1	1	1	1				

In the next step, we write the flip-flop inputs in terms of the external inputs and flip-flop outputs as in

$$D = x \oplus Q_a'(t) \qquad T = y' + Q_a'(t).$$

Using the above equations, we fill the D and T columns of the state table as in Table 1.12.

Table 1.12 Filled D and T columns

x	y	$Q_a(t)$	$Q_b(t)$	D	T	$Q_a(t+1)$	$Q_b(t+1)$
0	0	0	0	0	1		
0	0	0	1	0	1		
0	0	1	0	0	1		
0	0	1	1	0	1		
0	1	0	0	0	1		
0	1	0	1	0	1		
0	1	1	0	0	0		
0	1	1	1	0	0		
1	0	0	0	1	1		
1	0	0	1	1	1		
1	0	1	0	0	1		
1	0	1	1	0	1		
1	1	0	0	1	1		
1	1	0	1	1	1		
1	1	1	0	0	0		
1	1	1	1	0	0		

Upon the application of the clock pulse, the output of the D flip-flop is calculated as $Q(t+1) = D$, and the output of the T flip-flop is calculated using $Q(t+1) = T \oplus Q(t)$ leading to Table 1.13.

Table 1.13 State table completed

x	y	$Q_a(t)$	$Q_b(t)$	D	T	$Q_a(t+1)$	$Q_b(t+1)$
0	0	0	0	0	1	0	1
0	0	0	1	0	1	0	0
0	0	1	0	0	1	0	1
0	0	1	1	0	1	0	0
0	1	0	0	0	1	0	1
0	1	0	1	0	1	0	0
0	1	1	0	0	0	0	0
0	1	1	1	0	0	0	1
1	0	0	0	1	1	1	1
1	0	0	1	1	1	1	0
1	0	1	0	0	1	0	1
1	0	1	1	0	1	0	0
1	1	0	0	1	1	1	1
1	1	0	1	1	1	1	0
1	1	1	0	0	0	0	0
1	1	1	1	0	0	0	1

The information available in a state table can be expressed using the state diagrams, i.e., a state table can be converted to a state diagram. In the next example, we illustrate this concept.

Example 1.26 The state table of a sequential circuit is given as in Table 1.14. Obtain the state diagram of this circuit using the state table.

Table 1.14 State table for Example 1.26

x	$Q_a(t)$	$Q_b(t)$	$Q_a(t+1)$	$Q_b(t+1)$	y
0	0	0	0	1	0
0	0	1	0	0	0
0	1	0	0	0	1
0	1	1	0	1	1
1	0	0	1	0	0
1	0	1	1	1	0
1	1	0	1	1	1
1	1	1	1	0	0

Solution 1.26 Let us construct the state diagram using the state table part-by-part. The shaded regions corresponding to $Q_a(t)Q_b(t) = 00$, $x = 0$ and $x = 1$ depicted on the left side of Fig. 1.77 can be shown by a state diagram as indicated on the right side of Fig. 1.77.

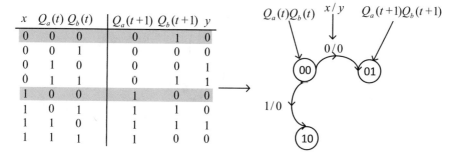

x	$Q_a(t)$	$Q_b(t)$	$Q_a(t+1)$	$Q_b(t+1)$	y
0	0	0	0	1	0
0	0	1	0	0	0
0	1	0	0	0	1
0	1	1	0	1	1
1	0	0	1	0	0
1	0	1	1	1	0
1	1	0	1	1	1
1	1	1	1	0	0

Fig. 1.77 State transitions when present state is 00

In a similar manner, the dark regions corresponding to $Q_a(t)Q_b(t) = 01$, $x = 0$ and $x = 1$ depicted on the left side of Fig. 1.77 can be shown by the state diagram on the right side of Fig. 1.78.

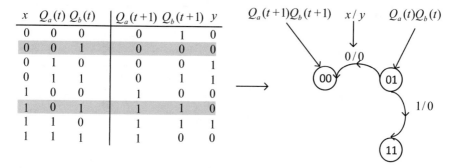

x	$Q_a(t)$	$Q_b(t)$	$Q_a(t+1)$	$Q_b(t+1)$	y
0	0	0	0	1	0
0	0	1	0	0	0
0	1	0	0	0	1
0	1	1	0	1	1
1	0	0	1	0	0
1	0	1	1	1	0
1	1	0	1	1	1
1	1	1	1	0	0

Fig. 1.78 State table to state diagram conversion operation

Proceeding in a similar manner we can construct the state table as in Fig. 1.79.

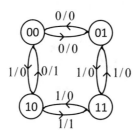

Fig. 1.79 Complete state diagram for Example 1.26

If we employ the letters for the states as $A = 00$, $B = 01$, $C = 10$, $D = 11$, we can draw the state diagram as in Fig. 1.80.

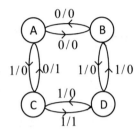

Fig. 1.80 State diagram with letter names

Example 1.27 Using the state table in Table 1.15, obtain the state diagram of the circuit.

Table 1.15 State table for Example 1.27

x	$Q_a(t)$	$Q_b(t)$	$Q_a(t+1)$	$Q_b(t+1)$
0	0	0	1	1
0	0	1	1	1
0	1	0	0	0
0	1	1	1	0
1	0	0	0	1
1	0	1	1	0
1	1	0	0	1
1	1	1	0	0

Solution 1.27 Using the information available in the state table, we can obtain the state diagram as in Fig. 1.81.

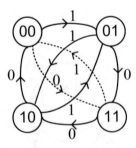

Fig. 1.81 State diagram of Table 1.15

Exercise   Using the state table given in Table 1.16, obtain the state diagram.

Table 1.16   State table for exercise

x	$Q_a(t)$	$Q_b(t)$	$Q_a(t+1)$	$Q_b(t+1)$
0	0	0	0	0
0	0	1	0	1
0	1	0	1	0
0	1	1	0	0
1	0	0	0	1
1	0	1	1	0
1	1	0	1	1
1	1	1	1	1

Example 1.28   In Fig. 1.82, a shift circuit with its initial value is depicted. Obtain the state diagram of this circuit.

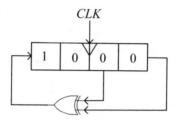

Fig. 1.82   Digital circuit with a shift register

Solution 1.28   Considering the operation of the shift register, we can construct the Moore state diagram of the circuit as in Fig. 1.83.

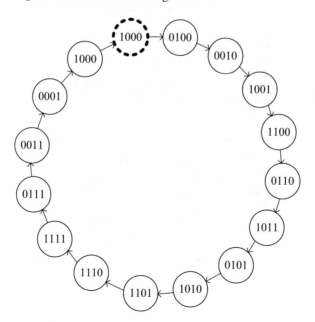

Fig. 1.83   State diagram of Example 1.28

Example 1.29 Let us try to find the state diagram of the convolutional encoder shown in Fig. 1.84.

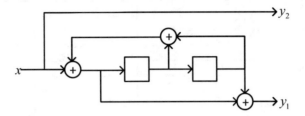

Fig. 1.84 Convolutional encoder

Since there are two cells in the convolutional encoder, the contents, i.e., outputs, of both memory cells can be one of the bit pairs 00, 01, 10, and 11. This means that there are four states of this convolutional encoder. Let us denote the pairs 00, 01, 10, and 11 by the symbols S_0, S_1, S_2, and S_3 respectively.

Initially, the contents of the registers are all zeros as depicted in Fig. 1.85.

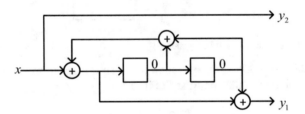

Fig. 1.85 Initial values of cells

When 0 is applied as external input, i.e., when $x = 0$, the outputs of the convolutional encoder can be calculated as shown in Fig. 1.86.

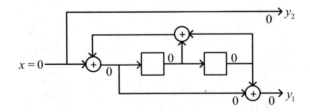

Fig. 1.86 Calculation of encoder outputs and next state for state S_0 and $x = 0$

And we can trace the behavior of the convolutional encoder by a state table or a state diagram as indicated in Fig. 1.87.

Present State S_i	Input x	Output y_2y_1	Next State S_i
$S_0 = 00$	0	00	

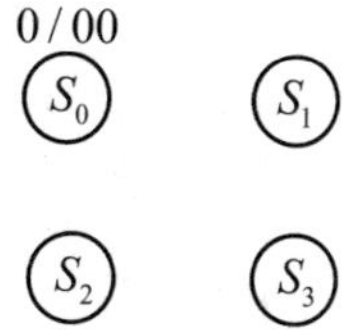

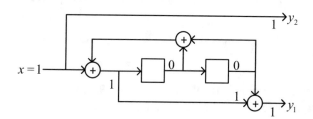

Fig. 1.87 State table and state diagram formation, step-1

Upon the application of clock pulse, the next state can be calculated using Fig. 1.86 for the input $x = 0$ as 00, i.e., S_0. We can update the state table and diagram shown in Fig. 1.87 as in Fig. 1.88.

Present State S_i	Input x	Output y_2y_1	Next State S_i
$S_0 = 00$	0	00	$S_0 = 00$

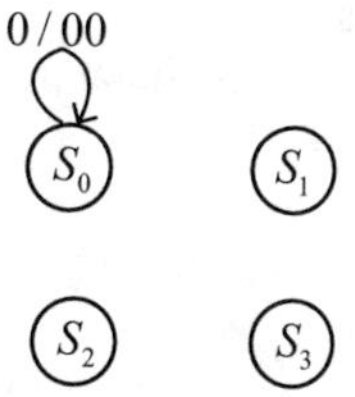

Fig. 1.88 State table and state diagram formation, step-2

When logic-1 is applied as external input, i.e., for $x = 1$, the outputs of the convolutional encoder can be calculated as shown in Fig. 1.89 when present state is S_0.

Fig. 1.89 Calculation of encoder outputs and next state for state S_0 and $x = 1$

The behavior of the convolutional encoder can be illustrated by a state table or a state diagram for present state S_0 and external input $x = 0$ as indicated in Fig. 1.90.

Present State S_i	Input x	Output y_2y_1	Next State S_i
$S_0 = 00$	0	00	$S_0 = 00$
$S_0 = 00$	1	11	

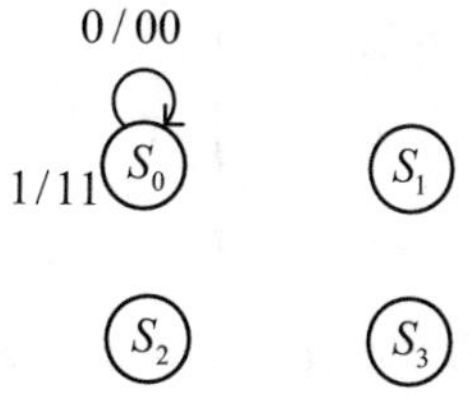

Fig. 1.90 State table and state diagram formation, step-3

For present state S_0 and external input $x = 1$, we can find the next state upon the application of clock pulse using Fig. 1.89 as 10, i.e., S_2. We can update the state table and diagram shown in Fig. 1.90 as in Fig. 1.91.

Present State	Input	Output	Next State
S_i	x	$y_2 y_1$	S_i
$S_0 = 00$	0	00	$S_0 = 00$
$S_0 = 00$	1	11	$S_2 = 10$

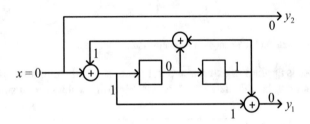

Fig. 1.91 State table and state diagram formation, step-4

For present state S_1 and external input $x = 0$, the output of the circuit can be found as $y_2 y_1 = 00$ as illustrated in Fig. 1.92, and upon clock pulse the next state happens to be $S_2 = 10$.

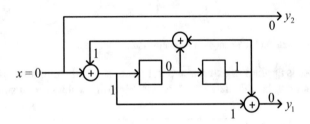

Fig. 1.92 Calculation of encoder outputs and next state for state S_1 and $x = 0$

The behavior of the circuit for the present state S_1 and external input $x = 0$ is expressed using state table and state diagram as in Fig. 1.93.

Present State	Input	Output	Next State
S_i	x	$y_2 y_1$	S_i
$S_0 = 00$	0	00	$S_0 = 00$
$S_0 = 00$	1	11	$S_2 = 10$
$S_1 = 01$	0	00	$S_2 = 10$

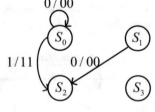

Fig. 1.93 State table and state diagram formation, step-5

For present state S_1 and external input $x = 1$, the output of the circuit can be found as $y_2y_1 = 11$ as illustrated in Fig. 1.94, and upon clock pulse the next state happens to be $S_0 = 00$.

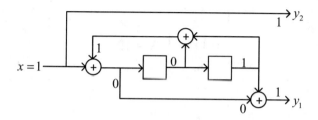

Fig. 1.94   Calculation of encoder outputs and next state for state S_1 and $x = 1$

The behavior of the circuit for the present state S_1 and external input $x = 1$ is expressed using state table and state diagram as in Fig. 1.95.

Present State	Input	Output	Next State
S_i	x	y_2y_1	S_i
$S_0 = 00$	0	00	$S_0 = 00$
$S_0 = 00$	1	11	$S_2 = 10$
$S_1 = 01$	0	00	$S_2 = 10$
$S_1 = 01$	1	11	$S_0 = 00$

Fig. 1.95   State table and state diagram formation, step-6

For present state S_2 and external input $x = 0$, the output of the circuit can be found as $y_2y_1 = 01$ as illustrated in Fig. 1.96, and upon clock pulse the next state happens to be $S_3 = 11$.

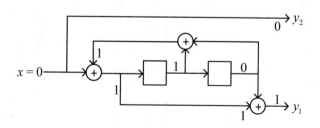

Fig. 1.96   Calculation of encoder outputs and next state for state S_2 and $x = 0$

The behavior of the circuit for the present state S_2 and external input $x = 0$ is expressed using state table and state diagram as in Fig. 1.97.

Present State S_i	Input x	Output y_2y_1	Next State S_i
$S_0 = 00$	0	00	$S_0 = 00$
$S_0 = 00$	1	11	$S_2 = 10$
$S_1 = 01$	0	00	$S_2 = 10$
$S_1 = 01$	1	11	$S_0 = 00$
$S_2 = 10$	0	01	$S_3 = 11$

Fig. 1.97 State table and state diagram formation, step-7

For present state S_2 and external input $x = 1$, the output of the circuit can be found as $y_2y_1 = 10$ as illustrated in Fig. 1.98, and upon clock pulse the next state happens to be $S_1 = 01$.

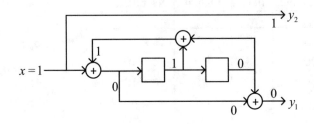

Fig. 1.98 Calculation of encoder outputs and next state for state S_2 and $x = 1$

The behavior of the circuit for the present state S_2 and external input $x = 1$ is expressed using state table and state diagram as in Fig. 1.99.

S_i	x	y_2y_1	S_i
$S_0 = 00$	0	00	$S_0 = 00$
$S_0 = 00$	1	11	$S_2 = 10$
$S_1 = 01$	0	00	$S_2 = 10$
$S_1 = 01$	1	11	$S_0 = 00$
$S_2 = 10$	0	01	$S_3 = 11$
$S_2 = 10$	1	10	$S_1 = 01$

Fig. 1.99 State table and state diagram formation, step-8

For present state S_3 and external input $x = 0$, the output of the circuit can be found as $y_2y_1 = 01$ as illustrated in Fig. 1.100, and upon clock pulse the next state happens to be $S_1 = 01$.

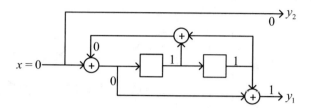

Fig. 1.100 Calculation of encoder outputs and next state for state S_3 and $x = 0$

The behavior of the circuit for the present state S_3 and external input $x = 0$ is expressed using state table and state diagram as in Fig. 1.101.

Present State	Input	Output	Next State
S_i	x	y_2y_1	S_i
$S_0 = 00$	0	00	$S_0 = 00$
$S_0 = 00$	1	11	$S_2 = 10$
$S_1 = 01$	0	00	$S_2 = 10$
$S_1 = 01$	1	11	$S_0 = 00$
$S_2 = 10$	0	01	$S_3 = 11$
$S_2 = 10$	1	10	$S_1 = 01$
$S_3 = 11$	0	01	$S_1 = 01$

Fig. 1.101 State table and state diagram formation, step-9

For present state S_3 and external input $x = 1$, the output of the circuit can be found as $y_2y_1 = 10$ as illustrated in Fig. 1.102, and upon clock pulse the next state happens to be $S_3 = 11$.

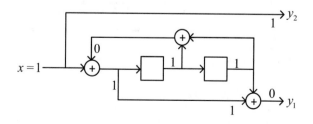

Fig. 1.102 Calculation of encoder outputs and next state for state S_3 and $x = 1$

The behavior of the circuit for the present state S_3 and external input $x = 1$ is expressed using state table and state diagram as in Fig. 1.103.

Present State S_i	Input x	Output $y_2 y_1$	Next State S_i
$S_0 = 00$	0	00	$S_0 = 00$
$S_0 = 00$	1	11	$S_2 = 10$
$S_1 = 01$	0	00	$S_2 = 10$
$S_1 = 01$	1	11	$S_0 = 00$
$S_2 = 10$	0	01	$S_3 = 11$
$S_2 = 10$	1	10	$S_1 = 01$
$S_3 = 11$	0	01	$S_1 = 01$
$S_3 = 11$	1	10	$S_3 = 11$

Fig. 1.103 State table and state diagram formation, step-10

The state diagram shown in Fig. 1.103 expresses the behavior of the circuit and it can be used to find the code-word for any input sequence.

Problems

1. Obtain the state diagram of the counter which repeats the count sequence 0, 3, 6, 7.
2. Obtain the Mealy and Moore state diagrams of the sequence detector which detects the sequence 0110. Overlapping patterns are allowed.
3. Obtain the state diagram of the convolutional encoder shown in Fig. P1.1.

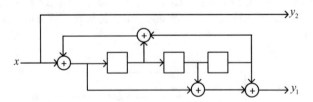

Fig. P1.1 Convolutional encoder

4. Convert the Mealy state diagram shown in Fig. P1.2 to Moore state diagram.

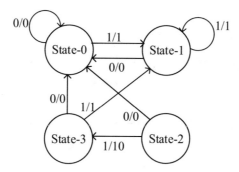

Fig. P1.2 Mealy state diagram for P4

5. Convert the Moore state diagram shown in Fig. P1.3 to Mealy state diagram.

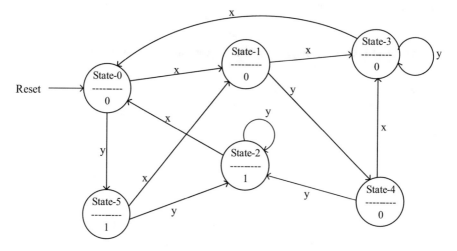

Fig. P1.3 Mealy state diagram for P5

6. Obtain the state diagram of the sequential circuit shown in Fig. P1.4.

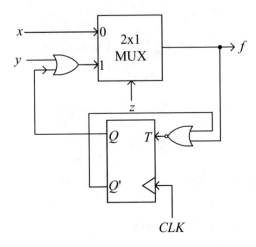

Fig. P1.4 Sequential circuit for P6

7. Draw the Mealy state diagram for RS232 transmission waveform.

VHDL Implementation of Finite State Machines and Practical Applications

2

In this chapter, we explain the VHDL implementation of finite state machines. We assume that the reader has the fundamental knowledge of VHDL programming. We do not aim to teach the fundamentals of VHDL programming in this chapter. We first explain the templates used in the implementation of state machines and then solve a variety of examples including some practical ones for the VHDL implementations. To be able to write VHDL programs for state diagrams, one should have the knowledge of primary concepts of state machines. For this reason, we suggest the reader to study Chap. 1 to have an idea of state machines before proceeding to this chapter.

2.1 Implementation of Finite State Machines in VHDL

A sequential circuit includes both memory elements, i.e., flip-flops, clock sources, and combinational logic units such as multiplexers, gates, decoders, encoders, etc. The general structure of a sequential logic circuit is depicted in Fig. 2.1.

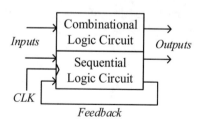

Fig. 2.1 General structure of a sequential circuit

A state machine can be implemented using sequential logic circuits. In other words, we can say that a state machine is nothing but a sequential logic circuit in practice.

© The Author(s), under exclusive license to Springer Nature Switzerland AG 2021
O. Gazi, A. Ç. Arlı, *State Machines using VHDL*,
https://doi.org/10.1007/978-3-030-61698-4_2

State machines can be divided into two main categories which are Moore and Mealy. In Moore state machines, the outputs of the circuit depend on the current values of the outputs of the memory elements. In other words, the outputs depend on the present states. A Moore state machine can have external inputs; however, the immediate changes of the external inputs have no immediate effects on the circuit outputs.

In Mealy state machines, the outputs of the circuit depend on the external inputs. When external inputs change, the outputs of the circuit may change before the completion of the current clock pulse.

A VHDL program written for the implementation of a state machine consists of mainly an entity part and two processes. One of the processes is written for the update of the current state upon the application of a clock pulse, and the other process is written for the determination of circuit outputs and deciding on the next state value. In addition, we can write a single process merging both processes, however, in this case, program readability decreases due to having many programming lines in a single process.

2.1.1   VHDL Implementation of Moore State Machines

The template for the entity and declarative part of the architecture unit of the Moore FSM is written in PR 2.1.

```
library ieee;
use ieee.std_logic_1164.all;

entity fsm_circuit is
  port(clk, rst: in std_logic;
         inp1, inp2,....,inpN: in data_type;
         outp1, outp2,....,outM: out data_type );
end entity;

architecture logic_flow of fsm_circuit is

type state is (st0, st1, st2, ...);
signal present_state, next_state: state;

begin
```

PR 2.1   Program 2.1

In PR 2.1, we define the circuit inputs and outputs in the entity unit. In the declarative part of the architecture, we introduce a new data type named **state** and using this new data type we declare two signal objects. Inside the body of the architecture unit, we have two processes. One of them is used for the update of the present state, and a template for this unit is given in PR 2.2.

```
-- Update of the present state
p1: process(clk, rst)
begin
  if(rst='1') then
    present_state<=st0;
  elsif(clk'event and clk='1') then
    present_state<=next_state;
  end if;
end process p1;
```

PR 2.2   Program 2.2

The sensitivity list of the process in PR 2.2 contains clock and reset signals, and at the rising edge of each clock pulse, present state update operation is performed. The template process unit for the determination of circuit outputs and next states is given in PR 2.3.

```
-- Circuit outputs and next states for Moore machines
p2: process(present_state, inp1, inp2,....)
begin

  case present_state is

    when st0 =>
      outp1<=oval1; outp2<=oval2; .... outpN<=ovalN;
      if(inp1=ival1) then
        next_state<=st1;
          :
      else
        next_state<=stM;
      end if;

    when st1 =>
      outp1<=oval3; outp2<=oval4; .... outpN<=ovalK;
      if(inp1=ival1) then
        next_state<=st3;
          :
      else
        next_state<=stM;
      end if;

    when ...
        :
  end case;

end process p2;
```

PR 2.3   Program 2.3

The sensitivity list of the process in PR 2.3 contains present state and input values. Whenever there is a change in the present state, the process "p2" in PR 2.3 is activated. This means that after the completion of process "p1" in PR 2.2, assuming that present state is updated, the process "p2" in PR 2.3 runs.

When all the parts are integrated, our template for Moore state machine happens to be as in PR 2.4.

```vhdl
library ieee;
use ieee.std_logic_1164.all;

entity fsm_circuit is
  port(clk, rst: in std_logic;
       inp1, inp2,....,inpN: in data_type;
       outp1, outp2,....,outM: out data_type );
end entity;
architecture logic_flow of fsm_circuit is
  type state is (st0, st1, st2,...);
  signal present_state, next_state: state;
begin
--- Update of the present state
  p1: process(clk, rst)
  begin
   if (rst='1') then
     present_state<=st0;
   elsif (clk'event and clk='1') then
     present_state<=next_state;
   end if;
  end process;
-- Circuit outputs and next state values for Moore machines
  p2: process(present_state, inp1, inp2,....)
  begin
   case present_state is
     when st0 =>
       outp1<=oval1; outp2<=oval2;....; outpN<=ovalN;
       if(inp1=ival1) then
         next_state<=st1;
            ⋮
       else
         next_state<=stM;
       end if;
     when st1 =>
       outp1<=oval3; outp2<=oval4;....; outpN<=ovalK;
       if(inp1=ival1) then
         next_state<=st3;
            ⋮
       else
         next_state<=stM;
       end if;
     when ...
        ⋮
   end case;
  end process;
end architecture;
```

PR 2.4 Program 2.4

It is important to note that in PR 2.4 the processes "p1" and "p2" runs in a sequential manner following the order of run "p1-p2-p1-p2...", and any change made on the value of "present_state" in process "p1" is seen by the process "p2" after the completion of process "p1".

2.1.2 VHDL Implementation of Mealy State Machines

The entity and declarative part of the architecture for the Mealy state machines are the same as that of PR 2.1 written for Moore state machines. Besides, the process for the update of the present state is also the same as PR 2.2 written for Moore state machines.

The only difference occurs in the implementation of the process written for the determination of circuit outputs and next states. In Mealy machines, the circuit outputs are determined considering the values of external inputs. For this reason, we should determine the outputs of the circuits after checking the values of external inputs by an **if** statement as in PR 2.5.

```
-- Circuit outputs and next states for Mealy machines
process(present_state, inp1, inp2,....)
begin

  case present_state is

    when st0 =>
    if(inp1=ival1) then
      outp1<=oval1; outp2<=oval2; .... outpN<=ovalN;
      next_state<=st1;
        ⋮
    else
      next_state<=stM;
    end if;

    when st1 =>
    if(inp1=ival2) then
      outp1<=oval3; outp2<=oval4; .... outpN<=ovalK;

      next_state<=st3;
        ⋮
    else
      next_state<=stM;
    end if;

    when ...
      ⋮
  end case;

end process;
```

PR 2.5 Program 2.5

When PR 2.5 is inspected, we see that the assignments to the output ports are performed after **if** statement.

2.2   Examples for VHDL Implementations of State Machines

In this section, we will provide a number of examples for the VHDL implementation of state machines.

2.2.1   Three-Bit Binary Counter in VHDL

We explain the implementation of 3-bit counter in VHDL via an example.

Example 2.1 Implement a 3-bit binary counter in VHDL.

Solution 2.1 Counters can be implemented using Moore state machines. Since they do not take external inputs other than clock and reset signals. For a 3-bit counter, we have 3 flip-flops, and hence 8 states in total.

Moore state diagram of the 3-bit binary counter can be drawn as in Fig. 2.2.

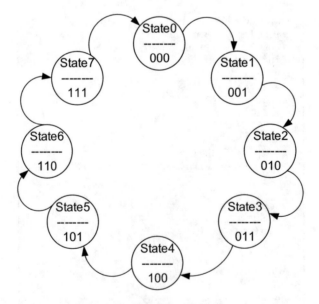

Fig. 2.2 Moore state diagram for 3-bit binary counter

We can write the entity unit and define the states in the declarative part of the architecture unit as in PR 2.6.

```
library ieee;
use ieee.std_logic_1164.all;

entity fsm_3bit_counter is
  port(clk, rst: in std_logic;
       outp: out std_logic_vector(2 downto 0));
end entity;

architecture logic_flow of fsm_3bit_counter is

  type state is (st0, st1, st2, st3, st4, st5, st6, st7);
  signal present_state, next_state: state;
begin
```

PR 2.6 Program 2.6

In the next step, we write the process for the present state update operation, and our program happens to be as in PR 2.7.

```
library ieee;
use ieee.std_logic_1164.all;

entity fsm_3bit_counter is
  port(clk, rst: in std_logic;
       outp: out std_logic_vector(2 downto 0));
end entity;

architecture logic_flow of fsm_3bit_counter is

  type state is (st0, st1, st2, st3, st4, st5, st6, st7);
  signal present_state, next_state: state;
begin

p1: process(clk, rst)
begin
  if(rst='1') then
    present_state<=st0;
  elsif(clk'event and clk='1') then
    present_state<=next_state;
  end if;
end process;
```

PR 2.7 Program 2.7

In the third step, we write the second process for the determination of circuit outputs and next states as in PR 2.8 where it is seen that the sensitivity list of the process contains the "present_state" signal object. This implies that the second process is activated only after the completion of the first process assuming that present state is updated in the first process.

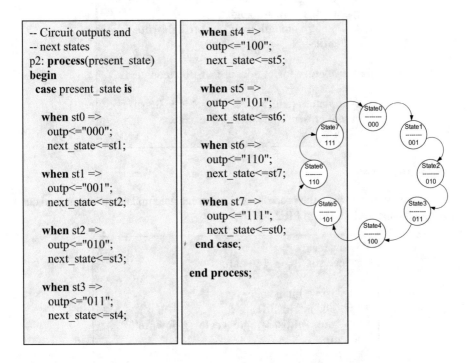

```
-- Circuit outputs and
-- next states
p2: process(present_state)
begin
 case present_state is

   when st0 =>
    outp<="000";
    next_state<=st1;

   when st1 =>
    outp<="001";
    next_state<=st2;

   when st2 =>
    outp<="010";
    next_state<=st3;

   when st3 =>
    outp<="011";
    next_state<=st4;

   when st4 =>
    outp<="100";
    next_state<=st5;

   when st5 =>
    outp<="101";
    next_state<=st6;

   when st6 =>
    outp<="110";
    next_state<=st7;

   when st7 =>
    outp<="111";
    next_state<=st0;
  end case;

end process;
```

PR 2.8 Program 2.8

When all the program units are integrated, our complete 3-bit counter program happens to be as in PR 2.9.

```
library ieee;
use ieee.std_logic_1164.all;

entity fsm_3bit_counter is
  port(clk, rst: in std_logic;
       outp: out std_logic_vector(2 downto 0));
end entity;
architecture logic_flow of fsm_3bit_counter is
  type state is (st0, st1, st2, st3, st4, st5, st6, st7);
  signal present_state, next_state: state;
begin
 p1: process(clk, rst)
 begin
  if(rst='1') then
    present_state<=st0;
   elsif(clk'event and clk='1') then
    present_state<=next_state;
   end if;
 end process;
-- Circuit outputs and next states
 p2: process(present_state)
 begin
  case present_state is
   when st0 =>
    outp<="000";
    next_state<=st1;
   when st1 =>
    outp<="001";
    next_state<=st2;
   when st2 =>
    outp<="010";
    next_state<=st3;
   when st3 =>
    outp<="011";
    next_state<=st4;
   when st4 =>
    outp<="100";
    next_state<=st5;
   when st5 =>
    outp<="101";
    next_state<=st6;
   when st6 =>
    outp<="110";
    next_state<=st7;
   when st7 =>
    outp<="111";
    next_state<=st0;
  end case;
 end process;
```

PR 2.9 Program 2.9

The VHDL implementation in PR 2.9 can be tested using the test-bench program given in PR 2.10.

```
library ieee;
use ieee.std_logic_1164.all;

entity fsm_3bit_counter_tb is
end;

architecture bench of fsm_3bit_counter_tb is

  component fsm_3bit_counter
    port(clk, rst: in std_logic;
         outp: out std_logic_vector(2 downto 0));
  end component;

  signal clk, rst: std_logic;
  signal outp: std_logic_vector(2 downto 0);

  constant clock_period: time:=10 ns;
  signal stop_the_clock: boolean;

begin

  pm: fsm_3bit_counter port map(clk   => clk,
                                rst   => rst,
                                outp => outp );
```

```
ps: process  --stimulus
begin
  rst<='1';
  wait for clock_period;
  rst<='0';

  wait for clock_period*7;

  stop_the_clock<=true;
  wait;
end process;

pc: process --clock generation
begin
  while not stop_the_clock loop
    clk<='0';
    wait for clock_period / 2;
    clk<='1';
    wait for clock_period / 2;
  end loop;
  wait;
end process;
end;
```

PR 2.10 Program 2.10

2.2.2 Counter State Machine Program Flow Analysis

In this section, we will analyze the flow of the VHDL implementation written for the counter state machine in Example 2.1, i.e., the previous example. Let us explain the operation of the PR 2.9 considering its state diagram. First, assume that reset signal is sent to FPGA device, then the process "p1" is activated, and the present state is initialized to "st0" as illustrated in Fig. 2.3.

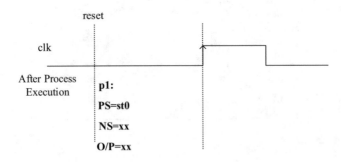

Fig. 2.3 Initialization at reset

When the execution of process "p1" finishes, the change on the signal object "present_state" can be seen by the other processes.

This is a critical point such that if you change the value of a signal object inside a process, the change is not immediately seen by the other program units. For the change to be seen by the other program units such as processes, the execution of the current process should be completed.

Since the sensitivity list of the process "p2" contains the signal object "present_state", after run of process "p1", the change on the signal object "present_state" triggers the process "p2", that is, after the run of process "p1", the process "p2" starts running. The process "p2" checks the value of the "present_state" and accordingly determines the circuit outputs and next state, as illustrated in Fig. 2.4.

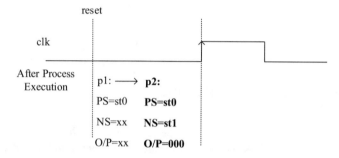

Fig. 2.4 "p2" activation after "p1"

After the application of "reset" signal, we consider the rising edge of the first incoming pulse at which the process "p1" runs, and when the process "p1" runs the present state is updated by the statement "present_state<=next_state". The update of the present state is illustrated in Fig. 2.5.

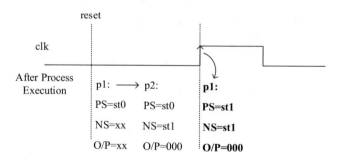

Fig. 2.5 "p1" activation at the first rising edge

From Fig. 2.5, it is seen that at the rising edge of the first clock pulse, the process "p1" is activated and present state value is changed to "st1"; however, next state and output values are kept the same. After the execution of process "p1", the process "p2" starts running, since the sensitivity list of the process "p2" contains the signal object "present_state". When the execution of process "p2" is completed, the next state and output values are recalculated as illustrated in Fig. 2.6.

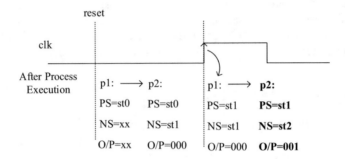

Fig. 2.6 "p2" activation after "p1" at the first rising edge

After the execution of process "p2", at the rising edge of the successor incoming pulse, the process "p1" is executed and the value of present state is updated as illustrated in Fig. 2.7.

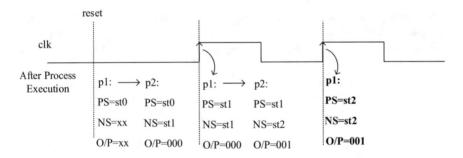

Fig. 2.7 "p1" activation at the second rising edge

After the run of process "p1", process "p2" is executed and circuit outputs and next state are determined as illustrated in Fig. 2.8.

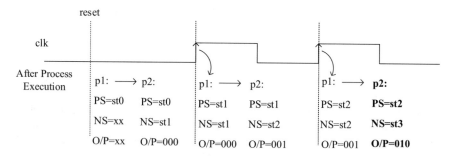

Fig. 2.8 "p2" activation after "p1" at the second rising edge

In a similar manner, considering the other rising edges of the other clock pulses, the complete execution of the state machine can be illustrated as in Fig. 2.9.

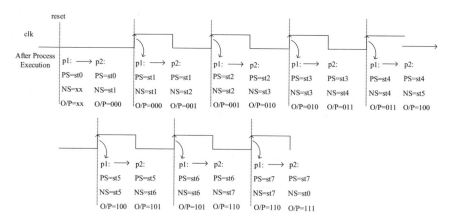

Fig. 2.9 "p1" and "p2" activations at rising edges

2.2.3 Predefined Encoding Types

The states of a logic machine can be represented by binary strings. Different representations result in different circuit syntheses. For this reason, it is important to assign the correct string to the states. The optimized assignments to the states are summarized as:

"Sequential": In this encoding method, the minimum number of bits is employed, and the states are encoded in ascending order of decimal values. For instance, A = "000" (= 0 decimal), B = "001" (= 1), C = "010" (= 2), D = "011" (= 3), and E = "100" (= 4).

"gray": In gray coding, the code-words of the adjacent states differ by exactly one bit. An M-bit "gray" code can represent 2M states. For instance, A = "000", B = "001", C = "011", D = "010", and E = "110".

"johnson": In johnson coding, M-bit encoding is used, and adjacent states differ by exactly one bit. For instance, A = "000", B = "100", C = "110", D = "111", and E = "011".

"one-hot": In one-hot coding, N bits, where N is the number of states, are used in enumeration type. Each code word contains only one different code bit (that is, all bits are '0', except one, or vice versa). For instance, A = "00001", B = "00010", C = "00100", D = "01000", and E = "10,000".

Although manual assignment of encoding types to states is possible, it is not much used in VHDL programming. Since FPGA development platforms, such as VIVADO, perform optimization for synthesis in which optimum encoding methods are used for states.

Example 2.2 In the code-segment below, the states are encoded using the "gray" coding approach.

```
type   states   is   (start,   first,   second,   delay,   success_1,
success_2);
   attribute enum_encoding: string;
   attribute enum_encoding of states: type is "gray";
```

2.2.4   Mealy State Diagram Implementation Example

In this section, we explain the VHDL implementation of a Mealy state diagram with an example.

Example 2.3 Implement the Mealy state machine whose state diagram is given in Fig. 2.10 where *x/y* indicates the *input/output* pair.

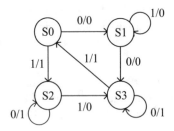

Fig. 2.10 Mealy state diagram

Solution 2.3 The entity and architecture declarative parts of the FSM can be written as in PR 2.11.

```
library ieee;
use ieee.std_logic_1164.all;

entity state_machine is
  port(clk: in std_logic;
       reset: in std_logic;
       inp: in std_logic;
       outp: out std_logic);
end state_machine;

architecture logic_flow of state_machine is
  type state is (st0, st1, st2, st3);  --type of state machine.
  signal present_state, next_state: state;
begin
```

PR 2.11 Program 2.11

The present state update part of the FSM can be written as in PR 2.12.

```
 --Present state update part
p1: process(clk, reset)
begin
  if(reset='1') then
    present_state<=st0;        --default state on
reset.
    elsif(rising_edge(clk)) then
      present_state<=next_state; --state change.
    end if;
```

PR 2.12 Program 2.12

The combinational logic part of the FSM can be written as in PR 2.13.

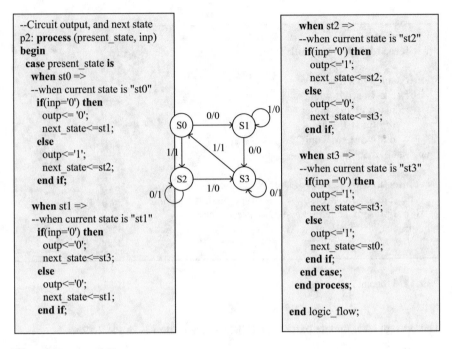

```
--Circuit output, and next state
p2: process (present_state, inp)
begin
 case present_state is
  when st0 =>
  --when current state is "st0"
   if(inp='0') then
    outp<= '0';
    next_state<=st1;
   else
    outp<='1';
    next_state<=st2;
   end if;

  when st1 =>
  --when current state is "st1"
   if(inp='0') then
    outp<='0';
    next_state<=st3;
   else
    outp<='0';
    next_state<=st1;
   end if;
```

```
  when st2 =>
  --when current state is "st2"
   if(inp='0') then
    outp<='1';
    next_state<=st2;
   else
    outp<='0';
    next_state<=st3;
   end if;

  when st3 =>
  --when current state is "st3"
   if(inp ='0') then
    outp<='1';
    next_state<=st3;
   else
    outp<='1';
    next_state<=st0;
   end if;
  end case;
 end process;

end logic_flow;
```

PR 2.13   Program 2.13

Combining all the program units, we get overall program as in PR 2.14.

```vhdl
library ieee;
use ieee.std_logic_1164.all;
entity state_machine is
  port(clk, reset: in std_logic;
       inp: in std_logic;
       outp: out std_logic);
end state_machine;

architecture logic_flow of state_machine is
  type state is (st0, st1, st2, st3);
  signal present_state, next_state: state;
begin

--present state update part
p1: process(clk, reset)
  begin
    if(reset='1') then
      present_state<= st0;   --default state on reset.
    elsif(rising_edge(clk)) then
      present_state<=next_state;   --state change.
    end if;
end process;
--Circuit outputs and next state
p2: process (present_state, inp)
  begin
    case present_state is

    when st0 =>  --when current state is "s0"
      if(inp='0') then
        outp<='0';
        next_state<=st1;
      else
        outp<='1';
        next_state<=st2;
      end if;

    when st1 => --when current state is "st1"
      if(inp='0') then
        outp<='0';
        next_state<=st3;
      else
        outp<='0';
        next_state<=st1;
      end if;

    when st2 => --when current state is "st2"
      if(inp='0') then
        outp<='1';
        next_state<=st2;
      else
        outp<='0';
        next_state<=st3;
      end if;

    when st3 => --when current state is "st3"
      if(inp='0') then
        outp<='1';
        next_state<=st3;
      else
        outp<='1';
        next_state<=st0;
      end if;
    end case;
end process;

end logic_flow;
```

PR 2.14 Program 2.14

The VHDL implementation in PR 2.14 can be tested using the test-bench program given in PR 2.15.

```
library ieee;
use ieee.std_logic_1164.all;

entity state_machine_tb is
end;

architecture bench of state_machine_tb is

  component state_machine
    port(clk, reset: in std_logic;
         inp: in std_logic;
         outp: out std_logic);
  end component;

  signal clk, reset: std_logic;
  signal inp: std_logic;
  signal outp: std_logic;

  constant clock_period: time:= 10 ns;
  signal stop_the_clock: boolean;

begin

  pm: state_machine port map (clk   => clk,
                              reset => reset,
                              inp   => inp,
                              outp  => outp );

ps: process   --stimulus
begin

    reset<='1';  reset<='0';

    inp<='1';
    wait for clock_period;

    inp<='1';
    wait for clock_period;

    inp<='1';
    wait for clock_period;

    inp<='0';
    wait for clock_period;

    stop_the_clock<=true;
    wait;
end process;

pc: process   --clock generation
begin
    while not stop_the_clock loop
      clk<='0';
      wait for clock_period / 2;
      clk<='1';
      wait for clock_period / 2;
    end loop;
    wait;
  end process;
end;
```

PR 2.15   Program 2.15

2.2.5   Parity Generator Implementation Example

In this section, we explain the VHDL implementation of parity generator with an example.

Example 2.4   Design a state machine for an even parity generator which is used for a sequence consisting of *N* bits. Implement your state machine in VHDL.

Solution 2.4 The state diagram of the even detector can be drawn as in Fig. 2.11.

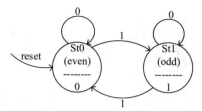

Fig. 2.11 State diagram of even parity generator

The entity and architecture declarative parts of the FSM can be written as in PR 2.16.

```
library ieee;
use ieee.std_logic_1164.all;

entity fsm_parity_generator is
  port(clk, reset: in std_logic;
       inp: in std_logic;
       parity: out std_logic);
end fsm_parity_generator;

architecture logic_flow of fsm_parity_generator is
  type state is (st0, st1);
  signal present_state, next_state: state;
begin
```

PR 2.16 Program 2.16

The present state update part of the FSM can be written as in PR 2.17.

```
p1: process(clk, reset)   --Present state update
begin
  if(reset='1') then
    present_state<=st0;   --default state on reset.
  elsif(rising_edge(clk)) then
    present_state<=next_state;   --state change.
  end if;
end process;
```

PR 2.17 Program 2.17

The process unit for the determination of circuit outputs and next state based on Moore state machine logic can be written as in PR 2.18.

```
-- Moore State Machine
p2: process (present_state, inp)
begin
  case present_state is
    when st0 =>   --when current state is "st0"
      parity<='0';
      if(inp ='0') then
        next_state<=st0;
      else
        next_state<=st1;
      end if;
    when st1 =>   --when current state is "st1"
      parity<=1';
      if(inp ='0') then
        next_state<=st1;
      else
        next_state<=st0;
      end if;
    end case;
  end process;
end logic_flow;
```

PR 2.18   Program 2.18

Combining all the program units, we get our overall program as in PR 2.19.

```
library ieee;
use ieee.std_logic_1164.all;
entity fsm_parity_generator is
  port(clk, reset: in std_logic;
       inp: in std_logic;
       parity: out std_logic);
end fsm_parity_generator;

architecture logic_flow of fsm_parity_generator is
  type state is (st0, st1);
  signal present_state, next_state: state;
begin
  --Present state update part
  p1: process (clk, reset)
  begin
    if(reset='1') then
      present_state<=st0;   --default state on reset.
    elsif(rising_edge(clk)) then
      present_state<=next_state;   --state change.
    end if;
  end process;
```

```
--- Moore State Machine
p2: process(present_state, inp)
begin
  case present_state is
    when st0 =>   --when current state is "st0"
      parity<='0';
      if(inp='0') then
        next_state<=st0;
      else
        next_state<=st1;
      end if;
    when st1 =>   --when current state is "st1"
      parity<='1';
      if(inp='0') then
        next_state<=st1;
      else
        next_state<=st0;
      end if;
    end case;
  end process;
end logic_flow;
```

PR 2.19   Program 2.19

The VHDL implementation in PR 2.19 can be tested using the test-bench in PR 2.20.

```
library ieee;
use ieee.std_logic_1164.all;

entity fsm_parity_generator_tb is
end;

architecture bench of fsm_parity_generator_tb
is

  component fsm_parity_generator
    port(clk, reset: in std_logic;
         inp: in std_logic;
         parity: out std_logic);
  end component;

  signal clk, reset: std_logic;
  signal inp: std_logic;
  signal parity: std_logic;

  constant clock_period: time:= 10 ns;
  signal stop_the_clock: boolean;

begin

pm: fsm_parity_generator
    port map (clk   => clk,
              reset => reset,
              inp   => inp,
              parity => parity );
```

```
ps: process   --stimulus
  begin

    reset<='1'; reset<='0';

    inp<='1';
    wait for clock_period;

    inp<='0';
    wait for clock_period;

    inp<='1';
    wait for clock_period;

    inp<='0';
    wait for clock_period;

    stop_the_clock<=true;
    wait;
  end process;

pc: process   --clock generation
  begin
    while not stop_the_clock loop
      clk<='0';
      wait for clock_period / 2;
      clk<='1';
      wait for clock_period / 2;
    end loop;
    wait;
  end process;
end;
```

PR 2.20 Program 2.20

2.2.6 Non-overlapping Sequence Detector Implementation Example

In this section, we explain the VHDL implementation of a non-overlapping sequence detector with an example.

Example 2.5 Design the state machine for non-overlapping sequence detector which detects the sequence 010. Implement your design in VHDL.

Solution 2.5 The state machine for the detection of the binary sequence 010 can be designed as in Fig. 2.12.

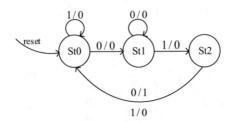

Fig. 2.12 State diagram of non-overlapping sequence detector

The entity and architecture declarative parts of the FSM presented in Fig. 2.12 can be written as in PR 2.21.

```
library ieee;
use ieee.std_logic_1164.all;

entity non_overlapping_detector is
  port(clk, reset: in std_logic;
       inp: in std_logic;
       outp: out std_logic);
end non_overlapping_detector;

architecture logic_flow of non_overlapping_detector is
  type state is (st0, st1, st2);
  signal present_state, next_state: state;
begin
```

PR 2.21   Program 2.21

The present state update part of the FSM can be written as in PR 2.22.

```
p1: process (clk, reset)   --Present state update
begin
if(reset='1') then
  present_state<=st0; --default state on reset.
elsif(rising_edge(clk)) then
  present_state<=next_state;   --state change.
  end if;
end process;
```

PR 2.22   Program 2.22

The process unit for the determination of circuit outputs and next state based on Mealy model can be written as in PR 2.23.

```
-- Mealy State Machine
p2: process (present_state, inp)
begin
 case present_state is
  when st0 =>
  --when current state is 'st0'
   if(inp='0') then
     outp<='0';
     next_state<=st1;
   else
     outp<='0';
     next_state<=st0;
   end if;

  when st1 =>
  --when current state is 'st1'
   if(inp='0') then
     outp<='0';
     next_state<=st1;
   else
     outp<='0';
     next_state<=st2;
   end if;
  when st2 =>
  --when current state is 'st2'
   if(inp='0') then
     outp<='1';
     next_state<=st0;
   else
     outp<='0';
     next_state<=st0;
   end if;
  end case;
 end process;
end logic_flow;
```

reset   St0   0/0   St1   1/0   St2
1/0   0/0
0/0
0/1
1/0

PR 2.23 Program 2.23

Combining all the program units, we get our overall program as in PR 2.24.

```
library ieee;
use ieee.std_logic_1164.all;
entity non_overlapping_detector is
  port(clk, reset: in std_logic;
       inp: in std_logic;
       outp: out std_logic);
end non_overlapping_detector;

architecture logic_flow of non_overlapping_detector
is
  type state is (st0, st1, st2);
  signal present_state, next_state: state;
begin
--Present state update part
p1: process(clk, reset)
  begin
    if(reset='1') then
      present_state<=st0;   --default state on reset.
    elsif(rising_edge(clk)) then
      present_state<=next_state;   --state change.
    end if;
end process;
-- Mealy State Machine
p2: process(present_state, inp)
  begin
    case present_state is
      when st0 =>  --when current state is "st0"
        if(inp ='0') then

      outp<='0';
      next_state<=st1;
    else
      outp<='0';
      next_state<=st0;
    end if;

    when st1 =>  -- current state is "st1"
      if(inp='0') then
        outp<='0';
        next_state<= st1;
      else
        outp<='0';
        next_state<=st2;
      end if;
    when st2 =>  -- current state is "st2"
      if(inp='0') then
        outp<='1';
        next_state<=st0;
      else
        outp<='0';
        next_state<=st0;
      end if;
    end case;
  end process;
end logic_flow;
```

PR 2.24 Program 2.24

2.2.7   Arbiter Implementation Example

In this section, we explain the VHDL implementation of an arbiter with an example.

Example 2.6 An arbiter is an electronic device which manages the access to shared resources. The task of the arbiter is explained in Fig. 2.13 where three peripherals P1, P2, and P3 use a common bus to access to a common resource. It is shown in Fig. 2.13 that the arbiter has the inputs r1, r2, r3, and outputs g1, g2, and g3.

The bus can be used only for one peripheral at a time. The peripheral PX, which appeals the use of the bus, issues a request to the arbiter, i.e., the signal ri is made "1", and if the bus is free, then the arbiter authorizes the peripheral PX to use the bus. For example, if P1 wants to use the bus, then it issues the signal r1='1', and if the bus is idle at the time of the request, then the arbiter authorizes the peripheral P1 to use the bus informing the peripheral making g1='1'.

If multiple requests are received at the same time by the arbiter, then the access is granted based on preestablished priorities if the line is idle, otherwise, the peripheral holding the bus continues with its process.

Considering the given information and assuming that the priorities are provided as P1 > P2 > P3, design a finite state machine and implement it in VHDL.

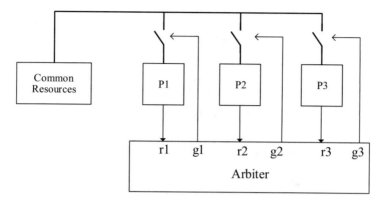

Fig. 2.13 Arbiter block diagram

Solution 2.6 In accordance with the provided information, the state diagram of the Moore machine for the arbiter can be drawn as in Fig. 2.14.

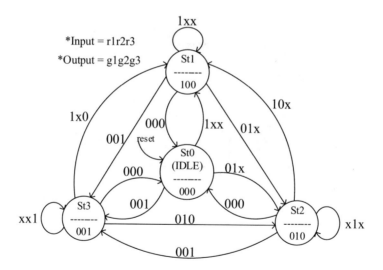

Fig. 2.14 Arbiter state diagram

The entity and declarative part of the architecture for the FSM shown in Fig. 2.14 can be written as in PR 2.25.

```
library ieee;
use ieee.std_logic_1164.all;

entity arbiter is
  port(clk, reset in std_logic;
        inp_r: in std_logic_vector (2 downto 0);
        outp_g: out std_logic_vector (2 downto 0));
end arbiter;

architecture logic_flow of arbiter is
  type state is (st0, st1, st2, st3);
  signal present_state, next_state: state;
begin
```

PR 2.25   Program 2.25

The process performing the state update of the FSM is written in PR 2.26.

```
 --Present state update part
p1: process(clk, reset)
begin
  if(reset='1') then
     present_state<=st0;        --default  state  on
reset.
   elsif(rising_edge(clk)) then
      present_state<=next_state;  --state change.
   end if;
```

PR 2.26   Program 2.26

The process for the determination of circuit outputs and next state based on Moore model can be written as in PR 2.27.

```
-- Circuit outputs and next state
p2: process(present_state, inp_r)
 begin
  case present_state is
   when st0 => --- idle state
    outp_g<="000";
    if(inp_r="000") then
     next_state<=st0;
    elsif(inp_r(2)='1') then
     next_state<=st1;
    elsif(inp_r(1)='1') then
     next_state<=st2;
    elsif(inp_r(0)='1') then
     next_state<=st3;
    end if;

   when st1 =>
    outp_g<="100";
    if(inp_r(2)= '1') then
     next_state<=st1;
    elsif(inp_r(1)='1') then
     next_state<=st2;
    elsif(inp_r(0)='1') then
     next_state<=st3;
    else
     next_state<=st0;
    end if;

   when st2 =>
    outp_g<="010";
    if(inp_r(1)='1') then
     next_state<=st2;
    elsif(inp_r(2)='1') then
     next_state<=st1;
    elsif(inp_r(0)='1') then
     next_state<=st3;
    else
     next_state<=st0;
    end if;

   when st3 =>
    outp_g<="001";
    if(inp_r(0)='1') then
     next_state<=st3;
    elsif(inp_r(2)='1') then
     next_state<=st1;
    elsif(inp_r(1)='1') then
     next_state<=st2;
    else
     next_state<=st0;
    end if;

  end case;
 end process;
end logic_flow;
```

PR 2.27 Program 2.27

Combining all the program units, we get the complete implementation in PR 2.28.

```vhdl
library ieee;
use ieee.std_logic_1164.all;

entity arbiter is
  port(clk, reset: in std_logic;
       inp_r: in std_logic_vector(2 downto 0);
       outp_g: out std_logic_vector(2 downto 0));
end arbiter;

architecture logic_flow of arbiter is
  type state is (st0, st1, st2, st3);
  signal present_state, next_state: state;
begin

--Present state update part
p1: process (clk, reset)
  begin
   if(reset='1') then
     present_state<=st0;  --default state on reset.
   elsif(rising_edge(clk)) then
     present_state<=next_state;  --state change.
   end if;
  end process;

-- Circuit outputs and next state
p2: process(present_state, inp_r)
  begin
   case present_state is
     when st0 => --- idle state
     outp_g<="000";
     if( inp_r="000") then
      next_state<=st0;
     elsif(inp_r(2)='1') then
      next_state<=st1;
     elsif(inp_r(1)='1') then
      next_state<=st2;
     elsif(inp_r(0)='1') then
      next_state<=st3;
     end if;

     when st1 =>
      outp_g<="100";
      if( inp_r(2)='1') then
       next_state<=st1;
      elsif(inp_r(1)='1') then
       next_state<=st2;
      elsif(inp_r(0)='1') then
       next_state<=st3;
      else
       next_state<=st0;
      end if;

     when st2 =>
      outp_g<="010";
      if( inp_r(1)='1') then
       next_state<=st2;
      elsif(inp_r(2)='1') then
       next_state<=st1;
      elsif(inp_r(0)='1') then
       next_state<=st3;
      else
       next_state<=st0;
      end if;

     when st3 =>
      outp_g<="001";
      if(inp_r(0)='1') then
       next_state<=st3;
      elsif(inp_r(2)='1') then
       next_state<=st1;
      elsif(inp_r(1)='1') then
       next_state<=st2;
      else
       next_state<=st0;
      end if;

   end case;
  end process;
end logic_flow;
```

PR 2.28 Program 2.28

2.2.8   VHDL Implementation of RS232 Asynchronous Serial Communication Protocol

Although in the current technology, most of the electronic devices use USB standard for serial communication, it is not confusing to see devices still employing RS232 serial communication. Besides, to comprehend the logic of serial communication and developing an interface between an electronic device having RS232 port and FPGA, we see it useful to implement the RS232 for FPGA devices.

Universal asynchronous receiver-transmitter, i.e., UART, is an electronic device consisting of some registers, and its main function is to convert parallel data to serial and serial data to parallel depending on the direction of communication. If the data is sent from the computer to an electronic device, the parallel data in computer registers are converted to serial via UART and it is transmitted through RS232 port.

RS232 serial communication is an asynchronous communication technique. An 8-bit ASCII code representing a symbol can be transmitted with RS232 standard in an asynchronous manner. The asynchronous transmitting waveform of RS232 standard is depicted in Fig. 2.15 where it is seen that least significant bit (LSB) is sent first and most significant bit (MSB) is sent last.

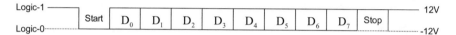

Fig. 2.15 RS232 transmitter waveform

A parity bit can be inserted between D_7 and "Stop" in the transmitted stream shown in Fig. 2.15.

Moore state diagram for the transmission waveform of Fig. 2.15 can be drawn as in Fig. 2.16.

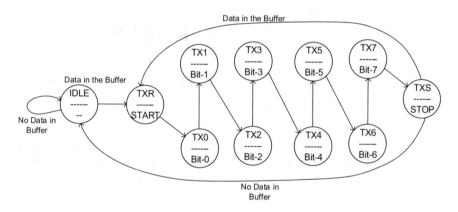

Fig. 2.16 RS232 transmitter state diagram

In Fig. 2.16, the state names are written above the dashed line, and the outputs are written below the dashed line.

Example 2.7 RS232 transmission waveform for the letter "J" whose ASCII code is 0x4A, i.e., 01001010 in binary, is depicted in Fig. 2.17.

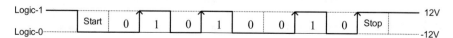

Fig. 2.17 RS232 transmitter waveform example

In Fig. 2.17, we should pay attention that the least significant bit is transmitted first.

2.2.8.1 VHDL Implementation of RS232 Transmitter

In this section, we explain the VHDL implementation of the RS232 serial communication protocol via examples. We first consider the transmission of data from FPGA to computer using RS232 communication protocol, i.e., first, we implement RS232 data transmission in VHDL, and in the successor examples, we implement RS232 receiver and RS232 transmitter-receiver structures in VHDL.

Example 2.8   For the RS232 asynchronous communication protocol parameters

- 9600 bit per second baud rate
- 8-bit data
- 1 stop bit
- No parity

write a VHDL program using finite state machine to transmit the character "A" represented by "01000001" in ASCII from FPGA to computer and observe it on a terminal program of windows. The system for this data transmission is depicted in Fig. 2.18.

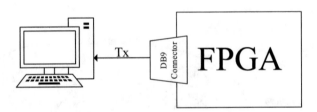

Fig. 2.18   FPGA to computer data transmission via RS232 protocol

Solution 2.8 Considering the asynchronous transmission waveform shown in Fig. 2.19 the Moore state diagram for the transmission scheme can be drawn as in Fig. 2.20 where "trig" signal is used to check the availability of new data to be transmitted.

We should keep in our mind that in this asynchronous transmission method, the least significant bit of the character "A" is sent first.

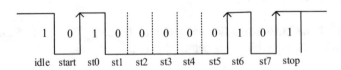

Fig. 2.19   RS232 transmission waveform for Example 2.8

The speed of the transmission is 9600 bits per second, and for this speed of transmission, we need a clock frequency of 9600 Hz. Since FPGA boards have clock frequencies in MHz range, we need a frequency divider to get 9600 Hz frequency.

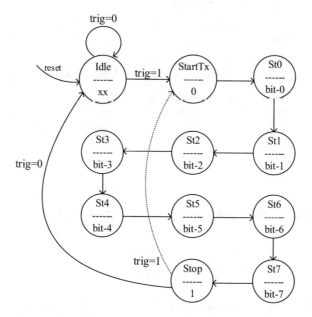

Fig. 2.20   State diagram of RS232 transmitter for Example 2.8

In PR 2.29, input and output ports and control signals are defined. State machine consists of 11 states, and states st0, st1, …, st7 are used for data transmission, "start" and "stop" states are considered to start and to stop the transmission, and "idle" state is used for port listening. To get the desired transmission frequency, we define a signal object "count" to be used in frequency division operation.

```
library ieee;
use ieee.std_logic_1164.all;

entity uart_tx is
  port(clk, rst, trig: in std_logic;
        data_to_send: in std_logic_vector(7 downto 0);
        tx: out std_logic);
end entity;

architecture logic_flow of uart_tx is
  type state is (idle, start_tx, st0, st1, st2, st3, st4, st5, st6, st7, stop);
  signal present_state, next_state: state:=idle;
  signal clk_9600Hz: std_logic:='0';
  signal count: positive range 1 to 5208:=1;
begin
```

PR 2.29 Program 2.29

It is known that using a counter with repeating sequence $1, \cdots, K-1$, we can obtain a frequency of

$$\frac{f}{2K}$$

from an FPGA clock source of frequency f Hz considering that a signal object is used for counter index.

According to this information, and considering the availability of 100 MHz FPGA clock, the counter parameter K can be calculated as

$$\frac{100 \times 10^6}{2K} = 9600 \rightarrow K \approx 5209 \tag{2.1}$$

which is used to initialize the "count" signal object defined in the declarative part of the architecture in PR 2.29. The clock divider, used to get a frequency of 9600 Hz, is implemented in process "cdiv" in PR 2.30.

```
library ieee;
use ieee.std_logic_1164.all;

entity uart_tx is
  port(clk, rst, trig: in std_logic;
       data_to_send: in std_logic_vector(7 downto 0);
       tx: out std_logic);
end entity;

architecture logic_flow of uart_tx is
  type state is (idle, start_tx, st0, st1, st2, st3, st4, st5, st6, st7, stop);
  signal present_state, next_state: state:=idle;
  signal clk_9600Hz: std_logic:='0';
  signal count: positive range 1 to 5209:=1;
begin

  cdiv: process(clk, rst)   -- clock divider (cdiv)
  begin
   if(rst='1') then
     count<=1;
   elsif(rising_edge(clk)) then
     count<=count+1;
     if(count=5209) then
       clk_9600Hz<=not clk_9600Hz;
       count<=1;
     end if;
   end if;
  end process;
```

PR 2.30 Program 2.30

Present state update operation at every rising edge of the artificially generated clock clk_9600Hz is implemented in PR 2.31 in process "p1" which follows the clock divider process.

```vhdl
library ieee;
use ieee.std_logic_1164.all;

entity uart_tx is
  port(clk, rst, trig: in std_logic;
       data_to_send: in std_logic_vector(7 downto 0);
       tx: out std_logic);
end entity;

architecture logic_flow of uart_tx is

  type state is (idle, start_tx, st0, st1, st2, st3, st4, st5, st6, st7, stop);
  signal present_state, next_state: state:=idle;
  signal clk_9600Hz: std_logic:='0';
  signal count: positive range 1 to 5209:=1;
begin

  cdiv: process(clk, rst)   -- clock divider (cdiv)
  begin
   if(rst='1') then
     count<=1;
   elsif(rising_edge(clk)) then
     count<=count+1;
     if(count=5209) then
       clk_9600Hz<=not clk_9600Hz;
       count<=1;
     end if;
   end if;
  end process;

  p1: process(clk_9600Hz, rst)
  begin
   if(rst='1') then
     present_state<=idle;
   elsif(rising_edge(clk_9600Hz )) then
     present_state<=next_state;
   end if;
  end process;
```

PR 2.31   Program 2.31

The complete program for the asynchronous data transmission from FPGA to computer including the process for determination of the next state is depicted in PR 2.32 where "clk_out" port signal is added to observe frequency divider result in simulations.

```vhdl
library ieee;
use ieee.std_logic_1164.all;

entity uart_tx is
  port(clk, rst, trig: in std_logic;
       data_to_send: in std_logic_vector(7 downto 0);
       tx, clk_out: out std_logic);
end entity;

architecture logic_flow of uart_tx is

  type state is (idle, start_tx, st0, st1, st2, st3, st4, st5, st6,
                 st7, stop);
  signal present_state, next_state: state:=idle;
  signal clk_9600Hz: std_logic:='0';
  signal count: positive range 1 to 5209:=1;
begin

  clk_out<=clk_9600Hz;

  cdiv: process(clk, rst) -- clock divider (cdiv)
  begin
    if(rst='1') then
      count<=1;
    elsif(rising_edge(clk)) then
      count<=count+1;
      if(count=5209) then
        clk_9600Hz<=not clk_9600Hz;
        count<=1;
      end if;
    end if;
  end process;

  p1: process(clk_9600Hz, rst)
  begin
    if(rst ='1') then
      present_state<=idle;
    elsif(rising_edge(clk_9600Hz )) then
      present_state<=next_state;
    end if;
  end process;
```

```vhdl
  p2: process(present_state, trig)
  begin
    case present_state is
      when idle =>
        tx<='1';
        if(trig='1') then
          next_state<=start_tx;
        else
          next_state<=idle;
        end if;
      when start_tx =>
        tx<='0';
        next_state<=st0;
      when st0 =>
        tx<=data_to_send(0);
        next_state<=st1;
      when st1=>
        tx<=data_to_send(1);
        next_state<=st2;
      when st2=>
        tx<=data_to_send(2);
        next_state<=st3;
      when st3=>
        tx<=data_to_send(3);
        next_state<=st4;
      when st4=>
        tx<=data_to_send(4);
        next_state<=st5;
      when st5=>
        tx<=data_to_send(5);
        next_state<=st6;
      when st6=>
        tx<=data_to_send(6);
        next_state<=st7;
      when st7=>
        tx<=data_to_send(7);
        next_state<=stop;
      when stop=>
        tx<='1';
        if(trig='0') then
          next_state<=idle;
        else
          next_state<=start_tx;
        end if;
    end case;
  end process;
end logic_flow;
```

PR 2.32   Program 2.32

A variety of terminal programs are available for computer to computer asynchronous serial communication. Tera Term VT is such a program which can be used to display the received characters, sent from FPGA side, on the computer screen as shown in Fig. 2.21.

The binary string used for the data transmission is 01000001 assigned to the parameter "data_to_send", and the trigger signal is activated four times.

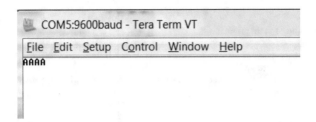

Fig. 2.21 Received data at Tera Term VT

The VHDL implementation in PR 2.32 can be tested using the test-bench given in PR 2.33.

```vhdl
library ieee;
use ieee.std_logic_1164.all;

entity uart_tx_tb is
end;

architecture bench of uart_tx_tb is

  component uart_tx
    port(clk, rst, trig: in std_logic;
         data_to_send: in std_logic_vector(7 downto 0);
         tx,clk_out: out std_logic);
  end component;

  signal clk, rst, trig: std_logic;
  signal data_to_send: std_logic_vector(7 downto 0);
  signal tx,clk_out: std_logic:='0';

  constant clock_period: time := 10 ns;
  signal stop_the_clock: boolean;

begin

  pm: uart_tx port map (
              clk => clk,
              rst => rst,
              trig => trig,
              data_to_send => data_to_send,
              tx=> tx,
              clk_out => clk_out);
```

```vhdl
ps: process   --stimulus
  begin

    rst<='1'; trig <= '0'; rst<='0';
    wait for clock_period*2*5208;
    trig <= '1';
    wait for clock_period*2*5208;

    data_to_send<="10010001";
    wait for clock_period*8*2*5208;

    stop_the_clock<=true;
    wait;
  end process;

  pc: process   --clock generation
  begin
    while not stop_the_clock loop
      clk<='0';
      wait for clock_period / 2;
      clk<='1';
      wait for clock_period / 2;
    end loop;
    wait;
  end process;

end;
```

PR 2.33 Program 2.33

2.2.8.2 VHDL Implementation of RS232 Receiver

In the previous example, we considered the transmission of data from FPGA to computer using RS232 protocol, i.e., FPGA is the transmitter and computer is the receiver. In the next example, we consider just the opposite scenario. That is, FPGA is the receiver and computer is the transmitter.

Example 2.9 Implement the RS232 receiver using a finite state machine on FPGA side using VHDL. That is, the computer side is the transmitter and the FPGA side is the receiver. The communication protocol parameters to be used are chosen as

- 9600 bit per second baud rate
- 8-bit data
- 1 stop bit
- No parity.

The received data on FPGA is displayed on LEDs of the FPGA board. The hardware setup is depicted in Fig. 2.22.

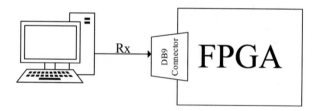

Fig. 2.22 Computer to FPGA data transmission via RS232 protocol

Solution 2.9 The state diagram for the VHDL implementation of the asynchronous receiver is depicted in Fig. 2.23 where it is seen that a control signal Rx is employed to start and stop the transmission process. Start and stop states are considered together not to lose any transmitted bit. Since the receiver side does not produce any output, the Moore states have only names, i.e., no outputs are written below the state names.

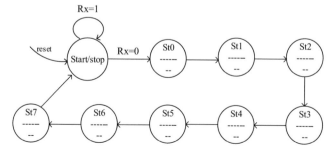

Fig. 2.23 RS232 receiver state diagram

In PR 2.34, input and output ports of the design and related states and signals are defined. We use "count" signal object in frequency divider to get the frequency of 9600 Hz as in previous example.

The state machine consists of nine states, and eight of them consisting of st0, st1, ..., st7 are used for data retrieval. The state "start_stop" state is employed to start or stop data acquisition.

```vhdl
library ieee;
use ieee.std_logic_1164.all;

entity uart_rx is
  port(clk, rst: in std_logic;
       outp: out std_logic_vector(7 downto 0);
       rx: in std_logic);
end entity;

architecture logic_flow of uart_rx is
  type state is (start_stop, st0, st1, st2, st3, st4, st5, st6, st7);
  signal present_state, next_state: state:=start_stop;
  signal clk_9600Hz: std_logic:='0';
  signal count: positive range 1 to 5208:=1;
begin
```

PR 2.34   Program 2.34

PR 2.35 includes the clock divider process to get the 9600 Hz clock frequency. The clock divider (frequency divider) process in PR 2.35 is the same as the one used in the previous example. The counter counts from 1 to 5208 according to the calculated value in (2.1), and it is assumed that FPGA has 100 MHz clock signal.

```
library ieee;
use ieee.std_logic_1164.all;

entity uart_rx is
  port(clk, rst: in std_logic;
       outp: out std_logic_vector(7 downto 0);
       rx: in std_logic);
end entity;

architecture logic_flow of uart_rx is
  type state is (start_stop, st0, st1, st2, st3, st4, st5, st6, st7);
  signal present_state, next_state: state:=start_stop;
  signal clk_9600Hz: std_logic:='0';
  signal count: positive range 1 to 5208:=1;

begin

  cdiv: process(clk, rst) -- cdiv means clock divider
  begin
   if(rst='1') then
     count<= 1;
   elsif(rising_edge(clk)) then
     count<=count+1;
     if(count=5208) then
       clk_9600Hz<=not clk_9600Hz;
       count<=1;
     end if;
    end if;
  end process;
```

PR 2.35   Program 2.35

The process "p1" used to update the present state is included in PR 2.36 where it is seen that the sensitivity list of the process contains the signals "clk_9600Hz" and "rst".

```vhdl
library ieee;
use ieee.std_logic_1164.all;

entity uart_rx is
  port(clk, rst: in std_logic;
       outp: out std_logic_vector(7 downto 0);
       rx: in std_logic);
end entity;

architecture logic_flow of uart_rx is
  type state is (start_stop, st0, st1, st2, st3, st4, st5, st6, st7);
  signal present_state, next_state: state:=start_stop;
  signal clk_9600Hz: std_logic:='0';
  signal count: positive range 1 to 5208:=1;
begin

  cdiv: process(clk, rst) -- cdiv means clock divider
  begin
   if(rst='1') then                PR 6-7
     count<=1;
   elsif(rising_edge(clk)) then
     count<=count+1;
    if(count=5208) then
      clk_9600Hz<=not clk_9600Hz;
      count<=1;
     end if;
    end if;
  end process;

  p1: process(clk_9600Hz, rst)
  begin
   if(rst = '1') then
     present_state<=start_stop;
   elsif(rising_edge(clk_9600Hz )) then
     present_state<=next_state;
    end if;
  end process;
```

PR 2.36 Program 2.36

Adding the process used for the determination of the next state values, we get the overall VHDL implementation as in PR 2.37 where we defined a new port signal "clk_out" to observe the "clk_9600Hz" signal in simulations.

```
library ieee;
use ieee.std_logic_1164.all;

entity uart_rx is
  port(clk, rst: in std_logic;
       outp: out std_logic_vector(7 downto 0);
       rx: in std_logic;
       clk_out: out std_logic);
end entity;

architecture logic_flow of uart_rx is
  type state is (start_stop, st0, st1, st2, st3, st4, st5,
                 st6, st7);
  signal present_state, next_state: state:=start_stop;
  signal clk_9600Hz: std_logic:='0';
  signal count: positive range 1 to 5208:=1;

begin
  clk_out<=clk_9600Hz;
  cdiv: process(clk, rst) -- cdiv means clock divider
  begin
    if(rst='1') then
      count<= 1;
    elsif(rising_edge(clk)) then
      count<=count+1;
      if(count=5208) then
        clk_9600Hz<=not clk_9600Hz;
        count<=1;
      end if;
    end if;
  end process;

  p1: process(clk_9600Hz, rst)
  begin
    if(rst = '1') then
      present_state<=start_stop;
    elsif(rising_edge(clk_9600Hz )) then
      present_state<=next_state;
    end if;
  end process;
```

```
p2: process(present_state, rx)
begin
  case present_state is
    when start_stop =>
      if(rx='1') then
        next_state<=start_stop;
      else
        next_state<=st0;
      end if;
    when st0 =>
      outp(0)<=rx;
      next_state<=st1;
    when st1=>
      outp(1)<=rx;
      next_state<=st2;
    when st2=>
      outp(2)<=rx;
      next_state<=st3;
    when st3=>
      outp(3)<=rx;
      next_state<=st4;
    when st4=>
      outp(4)<=rx;
      next_state<=st5;
    when st5=>
      outp(5)<=rx;
      next_state<=st6;
    when st6=>
      outp(6)<=rx;
      next_state<=st7;
    when st7=>
      outp(7)<=rx;
      next_state<=start_stop;
  end case;
end process;

end logic_flow;
```

PR 2.37 Program 2.37

The VHDL program in PR 2.37 can be tested using the test-bench program in PR 2.38.

```vhdl
library ieee;
use ieee.std_logic_1164.all;

entity uart_rx_tb is
end;

architecture bench of uart_rx_tb is

  component uart_rx
    port(clk, rst: in std_logic;
         outp: out std_logic_vector(7 downto 0);
         rx: in std_logic;
         clk_out: out std_logic);
  end component;

  signal clk, rst: std_logic;
  signal outp: std_logic_vector(7 downto 0);
  signal rx: std_logic;
  signal clk_out: std_logic;

  constant clock_period: time:=10 ns;
  signal stop_the_clock: boolean;

begin

pm: uart_rx port map(clk       => clk,
                     rst       => rst,
                     outp      => outp,
                     rx        => rx,
                     clk_out   => clk_out);

ps: process   --stimulus
begin

  rst<='1'; rst<='0'; rx<='1';
  wait for clock_period*2*5208;

  rx<='0';
  wait for clock_period*2*5208;

  rx<='1';
  wait for clock_period*2*5208;

  rx<='0';
  wait for clock_period*2*5208;

  rx<='1';
  wait for clock_period*2*5208;

  rx<='0';
  wait for clock_period*2*5208;

  rx<='1';
  wait for clock_period*2*5208;

  rx<='0';
  wait for clock_period*2*5208;

  rx<='1';
  wait for clock_period*2*5208;

  rx<='0';
  wait for clock_period*2*5208;

  stop_the_clock<=true;
  wait;
end process;

pc: process   --clock generation
begin
  while not stop_the_clock loop
    clk<='0';
    wait for clock_period / 2;
    clk<='1';
    wait for clock_period / 2;
  end loop;
  wait;
end process;

end;
```

PR 2.38   Program 2.38

2.2.8.3 VHDL Implementation of RS232 Transceiver

In the previous two examples, VHDL realizations of RS232 transmitter and receiver units were made. In the next example, we combine RS232 transmitter and receiver VHDL codes in the main program. For this purpose, we add new parameters to the transmit and receive programs and use "components" to utilize the transmit and receive programs in main program.

Example 2.10 In this example, both transmit and receive VHDL codes are utilized in a main VHDL program. We consider the scenario such that the serial data sent by computer is received by FPGA, and it is sent back to the computer, and on the terminal of the computer, the received character is displayed, i.e., the transmission scenario can be modeled as

$$TX_{Computer} \rightarrow RX_{FPGA} \rightarrow TX_{FPGA} \rightarrow RX_{Computer}.$$

We use components in main program. We take the serial communication protocol parameters as

- 9600 bit per second baud rate
- 8-bit data
- 1 stop bit
- No parity.

The character, sent from PC keyboard, is seen on a serial terminal window. In this scenario, FPGA can be considered as a bouncer. The hardware interfacing is depicted in Fig. 2.24 where it is seen that the FPGA part first receives the character sent by the computer side, and then it bounces it back to the computer, i.e., it sends it back to the computer side.

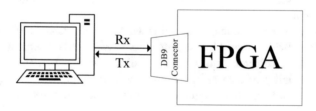

Fig. 2.24 RS232 transceiver for FPGA

Solution 2.10 The implementation consists of three units which are receiver, transmitter, and clock divider. The block diagram of the design is depicted in Fig. 2.25.

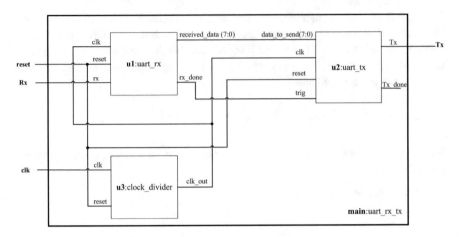

Fig. 2.25 RS232 transceiver block diagram

Transmitter and receiver blocks have additional flags denoted by "rx_done" and "tx_done" considering the transmitter and receiver programs of the previous two examples.

When a character is received, "rx_done" flag is used to indicate that the retrieval was successfully completed, i.e., "rx_done" is made "1", and the transmitter starts sending the received character. "tx_done" flag has no connection in this example, and it can be used for successful transmission of the transmission for improved implementations.

Since the same communication speed is used for both transmitter and receiver, only one clock divider is used in a separate block. We need two separate finite state machines for the receiver and transmitter as indicated in Fig. 2.26 where Rx=0 indicates the availability of data at the receiver port. It should not be forgotten that FPGA first performs the data acquisition operation, then sends the received data to the computer.

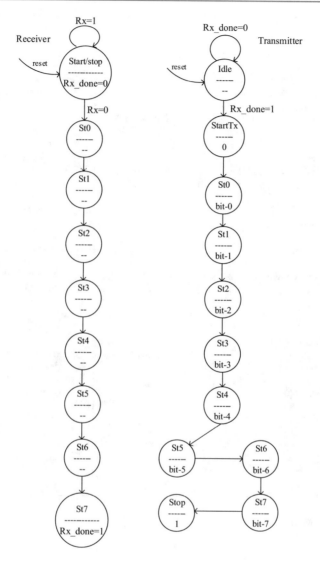

Fig. 2.26 RS232 transceiver state diagram

We write VHDL programs for receiver, transmitter, and clock divider separately, and use these programs in the main VHDL program using "components" for overall implementation.

First, the VHDL code for the receiver is written as shown in PR 2.39 where different from the previous examples, no process for clock division operation is used. "rx_done" flag is used for control purposes such that "rx=0" indicates the availability of data at the receiver port, and "rx_done=1" indicates the successful completion of the data acquisition.

```vhdl
library ieee;
use ieee.std_logic_1164.all;

entity uart_rx is
  port(clk, reset: in std_logic;
       rx_done: out std_logic;
       rx: in std_logic;
       received_data: out std_logic_vector(7 downto 0));
end uart_rx;

architecture logic_flow of uart_rx is
  type state is (start_stop, st0, st1, st2, st3, st4, st5, st6, st7);
  signal present_state, next_state: state:=start_stop;

begin
 p1: process(clk, reset)
 begin
  if(reset='1') then
    present_state<=start_stop;
  elsif(rising_edge(clk)) then
    present_state<=next_state;
  end if;
 end process;

 p2: process(present_state,rx)
 begin
   case present_state
    when start_stop =>
    rx_done<='0';
   if(rx='1') then
     next_state<=start_stop;
   else
     next_state<=st0;
   end if;
```

```vhdl
   when st0 =>
     received_data(0)<=rx;
     rx_done<='0';
     next_state<=st1;
   when st1=>
     received_data(1)<=rx;
     rx_done<='0';
     next_state<=st2;
   when st2=>
     received_data(2)<=rx;
     rx_done<='0';
     next_state<=st3;
   when st3=>
     received_data(3)<=rx;
     rx_done<='0';
     next_state<=st4;
   when st4=>
     received_data(4)<=rx;
     next_state<=st5;
     rx_done<='0';
   when st5=>
     received_data(5)<=rx;
     next_state<=st6;
     rx_done<='0';
   when st6=>
     received_data(6)<=rx;
     next_state<=st7;
     rx_done<='0';
   when st7=>
     received_data(7)<=rx;
     rx_done<='1';
     next_state<=start_stop;
   end case;
 end process;
end logic_flow;
```

PR 2.39 Program 2.39

Second, the VHDL code for the transmitter is written as shown in PR 2.40 where different from previous examples, no process for clock division operation is used, and "tx_done" flag is used to indicate the completion of the transmission.

```
library ieee;
use ieee.std_logic_1164.all;

entity uart_tx is
  port(clk, rst, trig: in std_logic;
       data_to_send: in std_logic_vector(7 downto 0);
       tx, tx_done: out std_logic);
end uart_tx;

architecture logic_flow of uart_tx is

  type state is (idle, start_tx, st0, st1, st2, st3, st4, st5,
                 st6, st7, stop);
  signal present_state, next_state: state:=idle;

begin

  p1: process(clk,rst)
  begin
   if(rst = '1') then
     present_state<=idle;
   elsif(rising_edge(clk)) then
     present_state<=next_state;
   end if;
  end process;

  p2: process(present_state, trig)
  begin
   case present_state is
    when idle =>
     tx<='1';
     tx_done<='0';
     if(trig='1') then
       next_state<=start_tx;
     else
       next_state<=idle;
     end if;
```

```
    when start_tx =>
     tx<='0';
     next_state<= st0;
    when st0 =>
     tx<=data_to_send(0);
     next_state<=st1;
    when st1=>
     tx<=data_to_send(1);
     next_state<=st2;
    when st2=>
     tx<=data_to_send(2);
     next_state<=st3;
    when st3=>
     tx<=data_to_send(3);
     next_state<=st4;
    when st4=>
     tx<=data_to_send(4);
     next_state<=st5;
    when st5=>
     tx<=data_to_send(5);
     next_state<=st6;
    when st6=>
     tx<= data_to_send(6);
     next_state<=st7;
    when st7=>
     tx<=data_to_send(7);
     next_state<=stop;
    when stop=>
     tx<='1';
     tx_done<='1';
     next_state<=idle;
   end case;
  end process;
end logic_flow;
```

PR 2.40   Program 2.40

In the third step, we implement the clock divider in a separate VHDL program, PR 2.41. Using clock (frequency) divider, we get 9600 Hz clock signal from a 100 MHz clock source.

```
library ieee;
use ieee.std_logic_1164.all;

entity clock_divider is
  port (clk, rst: in std_logic;
            clk_out: out std_logic);
end clock_divider;

architecture logic_flow of clock_divider is
  signal temp_clk_out: std_logic:='0';
  signal count: positive range 1 to 5208:=1;
begin
  process(clk, rst)
  begin
   if(rst='1') then
     count<=1;
   elsif(rising_edge(clk)) then
     count<=count+1;
     if(count=5208) then
       temp_clk_out<=not temp_clk_out;
       count<=1;
     end if;
    end if;
   end process;
   clk_out<=temp_clk_out;
end logic_flow;
```

PR 2.41 Program 2.41

Lastly, we write the main program as in PR 2.42 where component declarations for receiver, transmitter, and clock divider units are made in the declarative part of the architecture. We first initiate the receiver unit using the **port map** command and then initiate the transmitter part using the **port map** command. Initiation of the clock divider is performed last.

However, keep in your mind that the sequence of initiations is not important, since all the lines in the body of the architecture unit are performed in parallel.

```
library ieee;
use ieee.std_logic_1164.all;

entity uart_rx_tx is
  port(clk, reset: in std_logic;
       rx: in std_logic;
       tx: out std_logic);
end uart_rx_tx;

architecture logic_flow of uart_rx_tx is

  signal data. std_logic_vector(7 downto 0);
  signal rx_done, tx_done: std_logic;
  signal clk_9600Hz: std_logic;
  signal received_data: std_logic_vector(7 downto 0);

  component uart_rx
    port(clk, rst: in std_logic;
         rx: in std_logic;
         received_data: out std_logic_vector(7 downto 0);
         rx_done: out std_logic);
  end component;

  component uart_tx
    port(clk, rst, trig: in std_logic;
         data_to_send: in std_logic_vector(7 downto 0);
         tx, tx_done: out std_logic);
  end component;

  component clock_divider
    port(clk, rst: in std_logic;
         clk_out: out std_logic);
  end component;

begin
  u1: uart_rx port map(clk_9600Hz, rst, rx, received_data, rx_done);
  u2: uart_tx port map(clk_9600Hz, rst, rx_done, received_data, tx, tx_done);
  u3: clock_divider port map(clk, rst, clk_9600Hz);

end logic_flow;
```

PR 2.42 Program 2.42

In the next example, we improve the structure of the RS232 receiver implemented in VHDL. For this purpose, we consider the use of a buffer in the receiver system.

2.2.9 VHDL Implementation of FIFO

In this section, we explain the implementation of First-in First-out (FIFO) buffer in VHDL via an example.

Example 2.11 Design a First-in First-out (FIFO) data structure that is used to buffer incoming data for a predefined size. Use generic data types for data length and depth of the FIFO.

Solution 2.11 FIFOs are an indispensable design component used for buffering data for continuous data flow. It can be considered as a kind memory unit with read and write capabilities. A FIFO has empty and full control flags for the starting and stopping of data flow. The block diagram of a typical FIFO is given in Fig. 2.27.

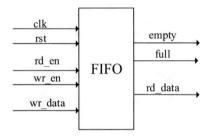

Fig. 2.27   FIFO buffer block diagram

The entity unit part of the architecture is given in PR 2.43 where port control and data signal objects for read and write operations are defined. Besides, generic definitions for input data and buffer sizes are used. For this example, although FIFO input data and buffer sizes are set to 4, it should be kept in mind that it is not an obligatory to choose FIFO data and buffer sizes the same.

```
library ieee;
use ieee.std_logic_1164.all;
use ieee.numeric_std.all;

entity fifo_module is
  generic(data_width: natural:= 4;
          fifo_depth: integer:= 4);

  port(clk, rst: in std_logic;
       wr_en: in  std_logic;
       wr_data: in  std_logic_vector(data_width-1 downto 0);
       full: out std_logic;
       rd_en: in  std_logic;
       rd_data: out std_logic_vector(data_width-1 downto 0);
       empty: out std_logic);
end fifo_module;
```

PR 2.43 Program 2.43

The signal objects to be used for the implementation of FIFO are defined in the declarative part of the architecture as in PR 2.44.

```
architecture logic_flow of fifo_module is

    type fifo_array_type is array (0 to fifo_depth-1) of std_logic_vector(data_width-1 downto 0);
    signal fifo_array: fifo_array_type:= (others => (others => '0'));
    signal wr_index: integer range 0 to fifo_depth-1:=0;
    signal rd_index: integer range 0 to fifo_depth-1:=0;
    signal fifo_line: integer range -1 to fifo_depth+1:=1;
    signal full_sig: std_logic;
    signal empty_sig: std_logic;

begin
```

PR 2.44   Program 2.44

The process written for FIFO read and write operations are given in PR 2.45.

```
process(clk, rst)
begin
    if(rst ='1') then
        fifo_line<=0;
        wr_index<=0;
        rd_index<=0;
    elsif(rising_edge(clk)) then

    if(wr_en='1' and rd_en='0') then
        fifo_array(wr_index)<=wr_data;
        fifo_line<=fifo_line + 1;
        if(wr_index=fifo_depth-1) then
            wr_index<=0;
        else
            wr_index<=wr_index + 1;
        end if;

    end if;

    if (wr_en='0' and rd_en='1') then
        fifo_line<=fifo_line-1;
        if(rd_index=fifo_depth-1) then
            rd_index<=0;
        else
            rd_index<=rd_index + 1;
        end if;
    end if;

end process;

rd_data<=fifo_array(rd_index) when wr_en='0' else x"0";
full_sig<='1' when fifo_line=fifo_depth else '0';
empty_sig<='1' when fifo_line=0 else '0';

full<=full_sig;
empty<=empty_sig;
end logic_flow;
```

PR 2.45   Program 2.45

Combining all the program units, we explained we get the overall implementation of PR 2.46.

```vhdl
library ieee;
use ieee.std_logic_1164.all;
use ieee.numeric_std.all;

entity fifo_module is
  generic(data_width: natural:= 4;
          fifo_depth: integer:= 4);

  port(clk, rst: in std_logic;
       wr_en: in  std_logic;
       wr_data: in std_logic_vector(data_width-1 downto 0);
       full: out std_logic;
       rd_en: in  std_logic;
       rd_data: out std_logic_vector(data_width-1 downto 0);
       empty: out std_logic);
end fifo_module;
architecture logic_flow of fifo_module is

  type fifo_array_type is array (0 to fifo_depth-1) of std_logic_vector(data_width-1 downto 0);
  signal fifo_array: fifo_array_type := (others => (others => '0'));
  signal wr_index: integer range 0 to fifo_depth-1:=0;
  signal rd_index: integer range 0 to fifo_depth -1:=0;
  signal fifo_line: integer range -1 to fifo_depth +1:=1;
  signal full_sig: std_logic;
  signal empty_sig: std_logic;

begin

  process(clk, rst)
  begin
   if (rst ='1') then
    fifo_line<=0;
    wr_index<=0;
    rd_index<=0;
   elsif(rising_edge(clk)) then

    if(wr_en='1' and rd_en='0') then
     fifo_array(wr_index)<=wr_data;
     fifo_line<=fifo_line + 1;
     if(wr_index=fifo_depth-1) then
       wr_index<=0;
     else
       wr_index<=wr_index + 1;
     end if;

    end if;
    if(wr_en='0' and rd_en='1') then
      fifo_line<=fifo_line-1;
      if(rd_index=fifo_depth-1) then
        rd_index<=0;
      else
        rd_index<=rd_index + 1;
      end if;
     end if;
    end if;

    end process;

    rd_data<=fifo_array(rd_index) when wr_en='0' else x"0";
    full_sig<='1' when fifo_line=fifo_depth else '0';
    empty_sig<='1' when fifo_line=0 else '0';

    full<=full_sig;
    empty<=empty_sig;
    end logic_flow;
```

PR 2.46   Program 2.46

2.2.10 VHDL Implementation of Buffered RS232 Receiver

In this section, we explain the VHDL implementation of buffered RS232 receiver via an example. We will employ components for our design. With this example, we aim how to make large implementations using component utility of VHDL programming.

Example 2.12 Implement the RS232 receiver system in FPGA using VHDL such that the data stream is received through Rx line and stored in a First-in First-out (FIFO) buffer. When FIFO is full, display the contents of FIFO on LEDs and empty the FIFO. Design your FIFO such that received data is displayed in every 0.5 s. New data stream can be received when FIFO is emptied. The communication protocol parameters used are

- 9600 bit per second baud rate
- 8-bit data
- 1 stop bit
- No parity.

Solution 2.12 The block diagram of the system is depicted in Fig. 2.28.

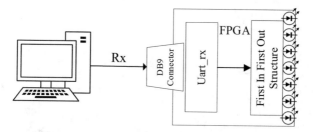

Fig. 2.28 Buffered RS232 receiver

The design consists of five units which are receiver, FIFO, LED controller, clock divider, and main program. In Fig. 2.29, the overall system to be implemented is illustrated.

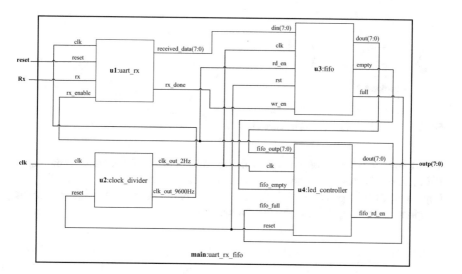

Fig. 2.29 Buffered RS232 receiver block diagram

Moore state machines for "uart_rx" and "led_controller" units are shown in Fig. 2.30. State machine design part for the RS232 receiver is a bit different from the previous examples.

The receiver is disabled when FIFO is full or while FIFO is emptying. The signals "rx_enable" or "fifo_read" are generated by "led_controller" unit.

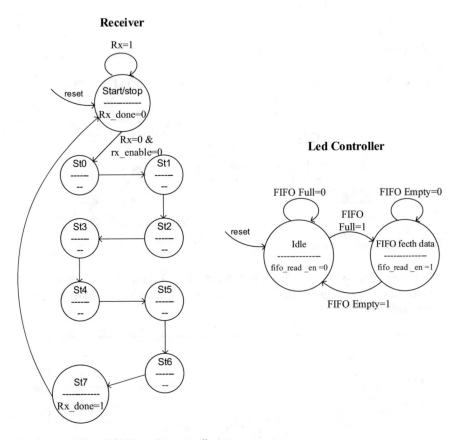

Fig. 2.30 Buffered RS232 receiver state diagram

VHDL implementation of the system in Fig. 2.30 consists of five units as indicated in Fig. 2.29. Four of these parts, receiver, FIFO, clock divider, and led controller, are implemented in separate VHDL files, and one main program uses these units as components and implements the overall system.

The VHDL program for receiver part is depicted in PR 2.47 wherein the entity part "rx_enable" and "rx_done" signals are defined for control purposes. The rest of the receiver program is similar to the receiver programs used in the previous examples.

```
library ieee;
use ieee.std_logic_1164.all;

entity uart_rx is
port(clk, rst: in std_logic;
     rx_enable, Rx: in std_logic;
     received_data: out std_logic_vector(7 downto 0);
     rx_done: out std_logic);
end uart_rx;

architecture logic_flow of uart_rx is
  type state is (start_stop, st0, st1, st2, st3, st4, st5, st6, st7);
  signal present_state, next_state: state:=start_stop;

begin
process(clk, rst)
begin
  if (rst = '1') then
    present_state<=start_stop;
  elsif(rising_edge(clk)) then
    present_state<=next_state;
  end if;
end process;

process(present_state, Rx)
begin
  case present_state is
    when start_stop =>
     if(Rx='1' and rx_enable='1' ) then
       next_state <= start_stop;
     elsif(Rx='0' and rx_enable='0') then
       next_state<=st0;
     else
       next_state<=start_stop;
     end if;
```

```
    when st0=>
      received_data(0)<=Rx;
      next_state<=st1;
    when st1=>
      received_data(1)<=Rx;
      next_state<=st2;
    when st2=>
      received_data(2)<=Rx;
      next_state<=st3;
    when st3=>
      received_data(3)<=Rx;
      next_state<=st4;
    when st4=>
      received_data(4)<=Rx;
      next_state<=st5;
    when st5=>
      received_data(5)<=Rx;
      next_state<=st6;
    when st6=>
      received_data(6)<=Rx;
      next_state<=st7;
    when st7=>
      received_data(7)<=Rx;
      if(rx_enable='0') then
        rx_done<= '1';
      else
        rx_done<= '0';
      end if;
      next_state<=start_stop;
    end case;
end process;
end logic_flow;
```

PR 2.47 Program 2.47

The led controller is implemented in PR 2.48.

```vhdl
library ieee;
use ieee.std_logic_1164.all;

entity led_controller is
  port(clk, rst: in std_logic;
       fifo_full, fifo_empty: in std_logic;
       fifo_outp: in std_logic_vector(7 downto 0);
       fifo_rd_en: out std_logic;
       dout: out std_logic_vector(7 downto 0));
end led_controller;

architecture logic_flow of led_controller is
  type state is (idle, data_out);
  signal present_state, next_state: state:=idle;
begin
  process(clk, rst)
  begin
    if (rst = '1') then
      present_state<= idle;
    elsif(rising_edge(clk)) then
      present_state<=next_state;
    end if;
  end process;

process(present_state, fifo_full, fifo_empty)
begin
  case present_state is
    when idle=>
      if(fifo_full='1') then
        next_state<=data_out;
        fifo_rd_en<='1';
      else
        fifo_rd_en<='0';
        next_state<= idle;
      end if;
    when data_out =>
      dout<=fifo_outp;
      if(fifo_empty='0') then
        fifo_rd_en<='1';
        next_state<=data_out;
      else
        fifo_rd_en<='0';
        next_state<=idle;
      end if;
  end case;
end process;
end logic_flow;
```

PR 2.48   Program 2.48

Two separate clock signals are generated. One of the clocks drives the "uart_rx" module whereas the other one controls with FIFO data flow and led controller operation.

Desired clock generations via frequency dividers are implemented in PR 2.49.

```
library ieee;
use ieee.std_logic_1164.all;

entity clock_divider is
  port(clk, rst: in std_logic;
        clk_out_9600Hz: out std_logic;
        clk_out_2Hz: out std_logic);
end clock_divider;

architecture logic_flow of clock_divider is
  signal temp_clk_out_9600Hz: std_logic:='0';
  signal temp_clk_out_2Hz: std_logic:='0';
  signal count1: positive range 1 to 5208:=1;
  signal count2: positive range 1 to 25000000:=1;
begin
  process(clk, rst)
  begin
   if(rst='1') then
     count1<=1;
    elsif (rising_edge(clk)) then
     count1<=count1+1;
     if (count1=5208) then
       temp_clk_out_9600Hz<=not temp_clk_out_9600Hz;
       count1<=1;
     end if;
    end if;
   end process;

  process(clk, rst)
  begin
   if(rst='1') then
     count2<=1;
    elsif(rising_edge(clk)) then
     count2<=count2+1;
     if(count2=25000000) then
       temp_clk_out_2Hz<=not temp_clk_out_2Hz;
       count2<=1;
     end if;
    end if;
   end process;

  clk_out_9600Hz<=temp_clk_out_9600Hz;
  clk_out_2Hz<=temp_clk_out_2Hz;
end logic_flow;
```

PR 2.49   Program 2.49

For the FIFO buffer implementation, we use the VHDL code of PR 2.46 which is written for Example 2.11. The main program unit written using components is depicted in PR 2.50.

```vhdl
library ieee;
use ieee.std_logic_1164.all;
entity uart_rx_fifo is
port(clk, rst, Rx: in std_logic;
     outp: out std_logic_vector(7 downto 0));
end uart_rx_fifo;

architecture logic_flow of uart_rx_fifo is
  signal data: std_logic_vector(7 downto 0);
  signal fifo_full, fifo_empty, rx_done, fifo_rd_en: std_logic;
  signal clk_9600Hz, clk_2Hz: std_logic;
  signal received_data: std_logic_vector(7 downto 0);
  signal fifo_outp: std_logic_vector(7 downto 0);
  signal led_out: std_logic_vector(7 downto 0);

  component uart_rx
    port(clk, rst :in std_logic;
         rx, rx_enable: in std_logic;
         received_data: out std_logic_vector(7 downto 0);
         rx_done: out std_logic);
  end component;

  component clock_divider is
    port(clk, rst: in std_logic;
         clk_out_9600Hz, clk_out_2Hz: out std_logic);
  end component;

  component fifo
    port(clk, rst, wr_en, rd_en: in std_logic;
         din: in std_logic_vector(7 downto 0);
         dout: out std_logic_vector(7 downto 0);
         full, empty: out std_logic );
  end component;

  component lcd_controller is
    port(clk, rst: in std_logic;
         fifo_full, fifo_empty: in std_logic;
         fifo_outp: in std_logic_vector(7 downto 0);
         fifo_rd_en: out std_logic;
         dout: out std_logic_vector(7 downto 0));
  end component;
begin
  u1: uart_rx port map(clk_9600Hz, rst, rx, fifo_rd_en, received_data, Rx_done);
  u2: clock_divider port map(clk, rst, clk_9600Hz, clk_2Hz);
  u3: fifo  port map (clk_2Hz, rst, received_data, rx_done, fifo_rd_en, fifo_outp, fifo_full, fifo_empty);
  u4: led_controller port map(clk_2Hz, rst, fifo_full, fifo_empty, fifo_outp, fifo_rd_en, outp);
end logic_flow;
```

PR 2.50   Program 2.50

Problems

1. Consider that the bit sequence "0101010" is transmitted via RS232 protocol. Draw the asynchronous timing waveform for the transmission of the given bit sequence.
2. Draw the timing waveform for the transmission of the character "A" by PS/2 protocol. The receiver is the FPGA and the sender is the computer.

3. Generate 100 Hz clock from 100 MHz FPGA clock source.
4. Draw the state diagram of the counter with the repeating sequence 0-4-7-6-8-8-0-3 and implement the counter in VHDL.
5. Draw the state diagram of the counter with the repeating sequence 2-6-7-8-1 such that between successive counts there is a time duration of 10 ms. Implement the counter in VHDL.
6. Implement the Mealy state machine shown in Fig. P2.1 in VHDL.

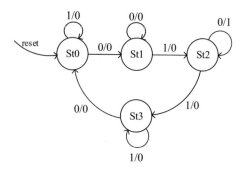

Fig. P2.1 Mealy state diagram for P6

7. Implement the Moore state machine shown in Fig. P2.2 in VHDL.

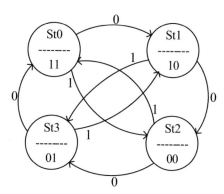

Fig. P2.2 Moore state diagram for P7

Timed Finite State Machines in VHDL

3

In this chapter, we explain timed state machines which can be considered as a general form of the state machines described in Chap. 2. In classical state machines, transition from one state to another occurs at every clock pulse. On the other hand, in timed state machines, a transition from one state to another occurs after several clock pulses, i.e., after a duration of time. Different amounts of time may be needed for the occurrence of transitions between different state pairs. Timed state machines are used in many practical applications. For instance, a traffic light controller used in daily life can be implemented using timed state machines.

3.1 Timed State Machine Models

In ordinary finite state machines, at each rising or falling edge of the clock pulse, a transition from one state to another one occurs. However, in some cases, those transitions from one state to another occur at multiples of clock period. The state machines with transitions occurring at multiples of clock period are called "timed state machines". The generic models for timed Moore and timed Mealy state machines are given in Figs. 3.1 and 3.2, respectively.

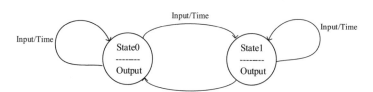

Fig. 3.1 Generic model for timed Moore state machine

© The Author(s), under exclusive license to Springer Nature Switzerland AG 2021
O. Gazi, A. Ç. Arlı, *State Machines using VHDL*,
https://doi.org/10.1007/978-3-030-61698-4_3

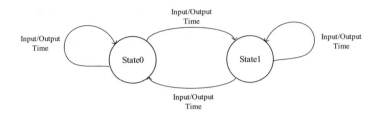

Fig. 3.2 Generic model for timed Mealy state machine

The "time" value placed along transitions show the number of clocks needed for the state transition to take place unless otherwise indicated.

Example 3.1 A timed Moore state machine is depicted in Fig. 3.3 where it is seen that if the present state is St0, when input is 1, after T_2 clocks, a transition from state St0 to St1 occurs, and at state St0 the circuit outputs are 01, and they are 10 for state St1.

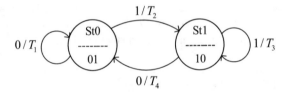

Fig. 3.3 A timed Moore state machine

3.2   VHDL Implementation of Timed Moore State Machines

The entity and declarative part of the architecture for the timed Moore FSM can be written as in PR 3.1.

```
library ieee;
use ieee.std_logic_1164.all;

entity fsm_circuit is
  port( clk, rst: in std_logic;
        inp1, inp2,..., inpN: in data_type;
        outp1, outp2,..., outM: out data_type );
end entity;

architecture logic_flow of fsm_circuit is

  type state is (st0, st1, st2,...);
  signal present_state, next_state: state;

  constant t1: natural:=t1_Value;
  constant t2: natural:=t2_Value;
    ⋮
  signal timer: natural range 0 to max_count;
  signal clk_count: natural range 0 to max_count;

begin
```

PR 3.1 Program 3.1

The newly added part of PR 3.1, when compared to the template of classical state machine, is shown inside a rounded rectangle.

Inside the body of the architecture unit, we have two processes. One of them is used for the update of the present state value, and a template for this unit is given in PR 3.2.

```
--- Update of the present state
p1: process(clk, rst)
begin
  if (rst='1') then
    present_state<=st0;
    clk_count<=0;
  elsif (clk'event and clk='1') then
    clk_count<= clk_count+1;
    if(clk_count=timer-1) then
      present_state<=next_state;
      clk_count<=0;
    end if;
  end if;
end process;
```

PR 3.2 Program 3.2

If we use variable object for "clk_count" rather than the signal object, the process "p1" is written in PR 3.3 where it is seen that "clk_count" is compared to "timer" rather than "timer-1". Since the update of the variable objects is immediate. On the other hand, the update of the signal objects is not immediate, and completion of the current process is required for the update of the signal objects.

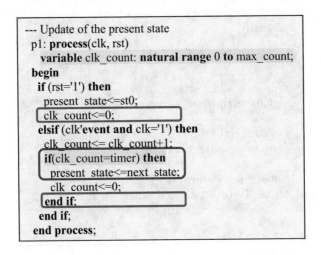

```
--- Update of the present state
p1: process(clk, rst)
   variable clk_count: natural range 0 to max_count;
begin
   if (rst='1') then
      present state<=st0;
      clk count<=0;
   elsif (clk'event and clk='1') then
      clk count<= clk count+1;
      if(clk count=timer) then
         present state<=next state;
      clk count<=0;
      end if;
   end if;
end process;
```

PR 3.3 Program 3.3

The template process unit for the determination of circuit outputs and next states is given in PR 3.4.

```
--- Circuit outputs and next states for Moore model
p2: process(present_state, inp1, inp2,...)
begin
  case present_state is
    when st0 =>
      outp1<=oval1;  outp2<=oval2;... ; outpN<=ovalN;
    if(inp1=ival1) then
      next_state<=st1;
      timer<=t1;
    elsif(inp1=ival2) then
      next_state<=st2;
      timer<=t2;
        ⋮
    else
      next_state<=stN;
      timer<=tm;
    end if;
    when st1 =>
      outp1<=oval3;  outp2<=oval4;... ; outpN<=ovalK;
    if(inp1=ival2) then
      next_state<=st1;
      timer<=t4;
    elsif(inp1=ival4) then
      next_state<=st2;
      timer<=t5;
        ⋮
    else
      next_state<=stN;
      timer<=tm;
    end if;
    when ...
        ⋮
  end case;
end process;
```

PR 3.4 Program 3.4

The sensitivity list of the process unit in PR 3.4 contains "present_state" signal object and input parameters.

It is also possible to have one process by merging the processes in PR 3.2 and PR 3.4. The process resulting from the merging of both processes is given in PR 3.5 where it is seen that sensitivity list of both processes are combined.

```
p1_p2: process(clk, rst, present_state, inp1, inp2,...)
begin
 if (rst='1') then
  present_state<=st0;
  clk_count<=0;
 elsif (clk'event and clk='1') then
  clk_count<= clk_count+1;
  if(clk_count=timer-1) then
   present_state<=next_state;
   clk_count<=0;
  end if;
 end if;
 case present_state is
  when st0 =>
   outp1<=oval1;  outp2<=oval2;...; outpN<=ovalN;
   if(inp1=ival1) then
    next_state<=st1;
    timer<=t1;
   elsif(inp1=ival2) then
    next_state<=st2;
    timer<=t2;
      ⋮
   else
    next_state<=stN;
    timer<=tm;
   end if;
  when st1 =>
   outp1<=oval3;  outp2<=oval4;...; outpN<=ovalK;
   if(inp1=ival2) then
    next_state<=st1;
    timer<=t4;
   elsif(inp1=ival4) then
    next_state<=st2;
    timer<=t5;
      ⋮
   else
    next_state<=stN;
    timer<=tm;
   end if;
  when ...
      ⋮
 end case;
end process;
```

PR 3.5 Program 3.5

When all the parts are combined, our template for Moore state machine happens to be as in PR 3.6.

```
library ieee;                                    --- Circuit outputs, next states and timer values for
use ieee.std_logic_1164.all;                     --- Moore model
                                                 p2: process(present_state, inp1, inp2,....)
entity fsm_circuit is                            begin
  port(clk, rst: in std_logic;                     case present_state is
        inp1, inp2,..., inpN: in  data_type;         when st0 =>
        outp1, outp2,..., outM: out  data_type );      outp1<=oval1;  outp2<=oval2;...;outpN<=ovalN;
end entity;                                            if(inp1=ival1) then
                                                         next_state<=st1;
architecture logic_flow of fsm_circuit is                timer<=t1;
                                                       elsif(inp1=ival2) then
  type state is (st0, st1, st2,...);                     next_state<=st2;
  signal present_state, next_state: state;               timer<=t2;
                                                              ⋮
  constant t1: natural:=t1_Value;                      else
  constant t2: natural:=t2_Value;                        next_state<=stN;
       ⋮                                                 timer<=tm;
  signal timer: natural range 0 to max_count;          end if;
  signal clk_count: natural range 0 to max_count;    when st1 =>
                                                         outp1<=oval3;  outp2<=oval4;...; outpN<=ovalK;
begin                                                    if(inp1=ival3) then
--- Update of the present state                            next_state<=st2;
p1: process(clk, rst)                                      timer<=t4;
begin                                                    elsif(inp1=ival4) then
  if (rst='1') then                                        next_state<=st2;
    present_state<=st0;                                    timer<=t5;
    clk_coun<=0;                                               ⋮
  elsif (clk'event and clk='1') then                     else
    clk_count<= clk_count+1;                               next_state<=stN;
    if(clk_count=timer-1) then                             timer<=tm;
      present_state<=next_state;                         end if;
      clk_count<=0;                                    when ...
    end if;                                                 ⋮
  end if;                                            end case;
end process;                                       end process;
                                                 end;
```

PR 3.6   Program 3.6

3.2.1   Timed Moore State Machine VHDL Implementation Example

In this section, we provide an example for the VHDL implementation of timed Moore state machine.

Example 3.2 Implement the timed Moore state machine shown in Fig. 3.4 in VHDL.

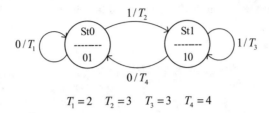

$$T_1 = 2 \quad T_2 = 3 \quad T_3 = 3 \quad T_4 = 4$$

Fig. 3.4 A timed Moore state diagram

Solution 3.2 The entity part and the declarative part of the architecture unit can be written as in PR 3.7.

```
library ieee;
use ieee.std_logic_1164.all;

entity timed_Moore_fsm is
  port(clk, rst: in std_logic;
        inp: in  std_logic;
        outp: out  std_logic_vector(1 downto 0) );
end entity;

architecture logic_flow of  timed_Moore_fsm is

type state is (st0, st1);
signal present_state, next_state: state;

constant t1: natural:=2;
constant t2: natural:=3;
constant t3: natural:=3;
constant t4: natural:=4;
constant max_count: natural:=4;
signal timer: natural range 0 to max_count;
signal clk_count: natural range 0 to max_count;
begin
```

PR 3.7 Program 3.7

The entity part of PR 3.7 contains clock, reset, input/output ports, and in the declarative part of the architecture of PR 3.7, state data type is defined and transition times between states are defined as constant objects. Besides, two more signal objects "timer" and "clk_count", to be used in processes, are introduced.

The state update process is given in PR 3.8.

```
p1: process(clk, rst)
begin
  if (rst='1') then
    present_state<=st0;
    clk_count<=0;
  elsif (clk'event and clk='1') then
    clk_count<=clk_count+1;
    if(clk_count=timer-1) then
      present_state<=next_state;
      clk_count<=0;
    end if;
  end if;
end process;
```

PR 3.8   Program 3.8

The process unit for the determination of circuit outputs and next states is given in PR 3.9.

```
--- Circuit outputs, next states
--- and timer values
p2: process(present_state, inp)
begin
  case present_state is
    when st0 =>
      outp<="01";
      if(inp='0') then
        next_state<=st0;
        timer<=t1;
      elsif(inp='1') then
        next_state<=st1;
        timer<=t2;
      end if;

    when st1 =>
      outp<="10";
      if(inp='0') then
        next_state<=st0;
        timer<=t4;
      elsif(inp='1') then
        next_state<=st1;
        timer<=t3;
      end if;
  end case;
end process;
```

PR 3.9   Program 3.9

Combining all the program units, we get the complete program as in PR 3.10.

```vhdl
library ieee;
use ieee.std_logic_1164.all;

entity timed_Moore_fsm is
 port(clk, rst: in std_logic;
      inp: in std_logic;
      outp: out std_logic_vector(1 downto 0) );
end entity;

architecture logic_flow of timed_Moore_fsm is

 type state is (st0, st1);
 signal present_state, next_state: state;

 constant t1: natural:=2;
 constant t2: natural:=3;
 constant t3: natural:=3;
 constant t4: natural:=4;
 constant max_count: natural:=4;
 signal timer: natural range 0 to max_count;
 signal clk_count: natural range 0 to max_count;
begin
 p1: process(clk, rst)
 begin
  if (rst='1') then
    present_state<=st0;
    clk_count<=0;
  elsif (clk'event and clk='1') then
    clk_count<=clk_count+1;
    if(clk_count=timer-1) then
      present_state<=next_state;
      clk_count<=0;
    end if;
  end if;
 end process;
```

```vhdl
--- Circuit outputs, next states
--- and timer values
 p2: process(present_state, inp)
 begin
  case present_state is

    when st0 =>
    outp<="01";
    if(inp='0') then
      next_state<=st0;
      timer<=t1;
    elsif(inp='1') then
      next_state<=st1;
      timer<=t2;
    end if;

    when st1 =>
    outp<="10";
    if(inp='0') then
      next_state<=st0;
      timer<=t4;
    elsif(inp='1') then
      next_state<=st1;
      timer<=t3;
    end if;
  end case;
 end process;

end logic_flow;
```

PR 3.10 Program 3.10

The VHDL program in PR 3.10 can be tested using the test-bench program in PR 3.11.

```
library ieee;
use ieee.std_logic_1164.all;

entity timed_Moore_fsm_tb is
end;

architecture bench of timed_Moore_fsm_tb is

 component timed_Moore_fsm
  port(clk, rst: in std_logic;
       inp: in std_logic;
       outp: out std_logic_vector(1 downto 0) );
 end component;

 signal clk, rst: std_logic;
 signal inp: std_logic;
 signal outp: std_logic_vector(1 downto 0) ;

 constant clock_period: time:=10 ns;
 signal stop_the_clock: boolean;

begin

 pm: timed_Moore_fsm port map(clk   => clk,
                              rst   => rst,
                              inp   => inp,
                              outp => outp);

 ps: process  --stimulus
  begin

  rst<='1'; rst<='0';

  inp<='1';
  wait for clock_period*3;
```

```
  inp<='1';
  wait for clock_period*3;

  inp<='0';
  wait for clock_period*4;

  inp<='0';
  wait for clock_period*2;

  inp<='1';
  wait for clock_period*3;

  inp<='1';
  wait for clock_period*3;

  inp<='0';
  wait for clock_period*4;

  stop_the_clock<=true;
  wait;
 end process;

 pc: process --clock generation
  begin
   while not stop_the_clock loop
    clk<='0';
    wait for clock_period / 2;
    clk<='1';
    wait for clock_period / 2;
   end loop;
   wait;
 end process;

end;
```

PR 3.11   Program 3.11

3.3   Analysis of the Timed Moore State Machine

In this section, we will analyze the VHDL implementation of the timed Moore state machine in PR 3.10. To trace of the flow of the VHDL program, we will assume that the input sequence is given as 101. We should keep in our mind that, if the value of a signal object is changed inside a process, this change is not seen outside the process unless the process execution finishes completely. Considering this important information, we can trace the execution of the timed state machines as follows.

Upon the application of the reset signal, the process "p1" works and present state (PS), next state (NS), output (O/P), clock count (clk_cnt), and timer values happen to be as in Fig. 3.5.

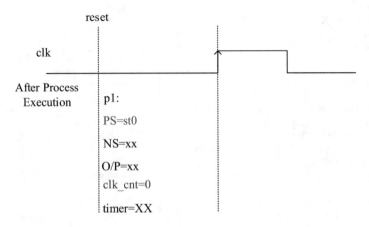

Fig. 3.5 Operation of "p1" after reset

The change of the present state triggers the second process "p2", and NS, O/P, and timer values are updated as in Fig. 3.6.

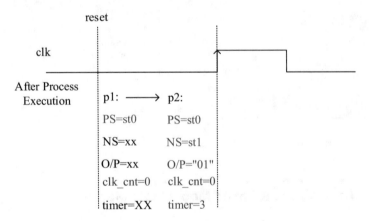

Fig. 3.6 Operations of "p1" and "p2" after reset

At the rising edge of the first clock pulse, "clk_cnt" value is incremented by 1, and the values of the other parameters stay the same as shown in Fig. 3.7.

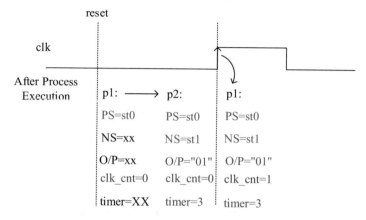

Fig. 3.7 Operation of "p1" at the first rising edge

At the rising edge of the second clock pulse, clk_cnt value is incremented by 1, and it is 2; however, this value is seen outside when the process execution finishes. For this reason, "**if(clk_count=timer-1)**" statement in process "p1" is not executed. The values of the other parameters stay the same as shown in Fig. 3.8.

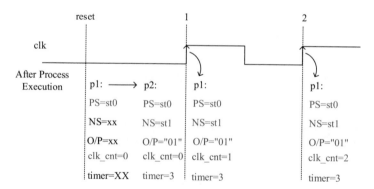

Fig. 3.8 Operation of "p1" at the second rising edge

At the rising edge of the third clock pulse, "clk_cnt" value is incremented by 1, and it becomes 3; however, this value is seen outside the process when the process execution finishes. Inside the process, its old value, i.e., 2 is seen. For this reason, inside process "p1", the Boolean expression in "**if(clk_count=timer-1)**" is evaluated as true, and "clk_cnt" value is initialized to 0, and present state value is updated as shown in Fig. 3.9.

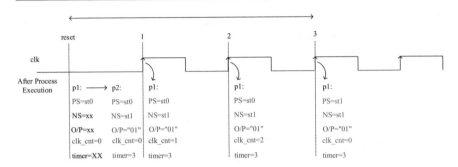

Fig. 3.9 Operation of "p1" at the third rising edge

Since the value of the present state is changed, the second process is triggered, as shown in Fig. 3.10, and output, timer values are updated as "O/P=10", "timer=4", and next state is determined as "st0" for input bit "0". It is seen from Fig. 3.10 that three clock cycles are required for the change of the present state.

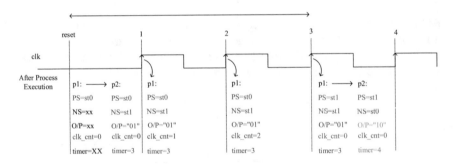

Fig. 3.10 Operations of "p1" and "p2" at the third rising edge

The trace of the program for the next four clock pulses is shown in Fig. 3.11 where we see that state change occurs after four clock pulses.

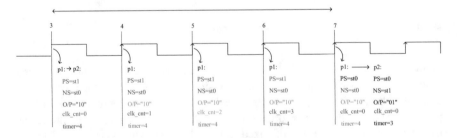

Fig. 3.11 Operations of "p1" and "p2" for the other rising edges

The trace of the program for the next three clock pulses is shown in Fig. 3.12 where we see that state change occurs after three clock pulses.

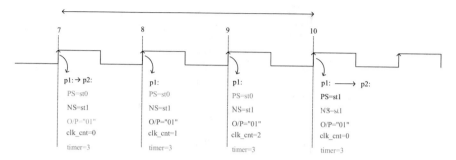

Fig. 3.12 Operations of "p1" and "p2" for the remaining rising edges

3.4   Seven-Segment Display as a Timed State Machine

In this section, we provide an example for the VHDL implementation seven-segment displays using Moore timed state machine.

Example 3.3 Write a VHDL program that drives a seven-segment display such that the digits on the seven-segment display are shown in a sequential manner with 1 s time durations. Assume that FPGA has 100 MHz clock generator.

Solution 3.3 The seven-segment display is an electronic unit used to display the digits 0, 1, …, 9. The symbolic representation of common-anode seven-segment (SS) display is depicted in Fig. 3.13.

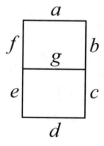

Fig. 3.13 SS display symbol

Table 3.1 BCD to SS conversion table

Digit	BCD code	Seven segment display code
	wxyz	abcdefg
0	0000	0000001
1	0001	1001111
2	0010	0010010
3	0011	0000110
4	0100	1001100
5	0101	0100100
6	0110	0100000
7	0111	0001111
8	1000	0000000
9	1001	0000100

The BCD codes representing digits and their corresponding seven-segment display codes are depicted in Table 3.1. Assume that FPGA's clock frequency is 100 MHz which corresponds to a period of $T = 10$ ns. To obtain 1 s time duration, we need 10^8 clock periods, since $10^8 T \rightarrow 1$ s. For this reason, the value of the constant object "t1" in PR 3.12 is initialized to 10^8.

The entity section and the declarative part of the architecture unit can be written as in PR 3.12.

```
library ieee;
use ieee.std_logic_1164.all;

entity ss_display_fsmis
  port(clk, rst: in std_logic;
       ssd: out std_logic_vector(6 downto 0));
end entity;

architecture logic_flow of ss_display_fsm is

  type state is (st0, st1, st2, st3, st4, st5, st6, st7, st8, st9);
  signal present_state, next_state: state;

  constant t1: natural:=100000000;
  constant max_count: natural:=100000000;
  signal timer: natural range 0 to max_count;
  signal clk_count: natural range 0 to max_count;
begin
```

PR 3.12 Program 3.12

The VHDL code for the initialization and update of the present state and counter value can be written as in PR 3.13.

```
p1: process(clk, rst)
begin
 if (rst='1') then
   present_state<=st0;
   clk_count<=0;
  elsif (clk'event and clk='1') then
   clk_count<=clk_count+1;
   if(clk_count=timer-1) then
    present_state<=next_state;
    clk_count<=0;
   end if;
  end if;
end process;
```

PR 3.13   Program 3.13

The calculation of circuit outputs, determination of next state, and update of the timer can be achieved as in PR 3.14.

```
p2: process(present_state)
begin
 case present_state is
  when st0 =>
   ssd<=st1;
   timer<=t1;
  when st1 =>
   ssd<="1001111";
   next_state<=st2;
   timer<=t1;
  when st2 =>
   ssd<="0010010";
   next_state<=st3;
   timer<=t1;
  when st3 =>
   ssd<="0000110";
   next_state<=st4;
   timer<=t1;
  when st4 =>
   ssd<="1001100";
   next_state<=st5;
   timer<=t1;
```

```
  when st5 =>
   ssd<="0100100";
   next_state<=st6;
   timer<=t1;
  when st6 =>
   ssd<="0100000";
   next_state<=st7;
   timer<=t1;
  when st7 =>
   ssd<="0001111";
   next_state<=st8;
   timer<=t1;
  when st8 =>
   ssd<="0000000";
   next_state<=st9;
   timer<=t1;
  when st9 =>
   ssd<="0000100";
   next_state<=st0;
   timer<=t1;
  end case;
 end process;
end logic_flow;
```

PR 3.14   Program 3.14

The complete program can be written as in PR 3.15.

```
library ieee;
use ieee.std_logic_1164.all;

entity ss_display_fsm is
  port(clk, rst: in std_logic;
       ssd: out std_logic_vector(6 downto 0));
end entity;

architecture logic_flow of ss_display_fsm is

  type state is (st0, st1, st2, st3, st4, st5, st6, st7, st8, st9);
  signal present_state, next_state: state;

  constant t1: natural:=100000000;
  constant max_count: natural:= 100000000;
  signal timer: natural range 0 to max_count;
  signal clk_count: natural range 0 to max_count;
begin
 p1: process(clk, rst)
 begin
   if (rst='1') then
     present_state<=st0;
     clk_count<=0;
   elsif (clk'event and clk='1') then
     clk_count<=clk_count+1;
```

```
   when st1 =>
     ssd<="1001111";
     next_state<=st2;
     timer<=t1;
   when st2 =>
     ssd<="0010010";
     next_state<=st3;
     timer<=t1;
   when st3 =>
     ssd<="0000110";
     next_state<=st4;
     timer<=t1;
   when st4 =>
     ssd<="1001100";
     next_state<=st5;
     timer<=t1;
   when st5 =>
     ssd<="0100100";
     next_state<=st6;
     timer<=t1;
   when st6 =>
     ssd<="0100000";
     next_state<=st7;
     timer<=t1;
   when st7 =>
```

PR 3.15 Program 3.15

To perform the simulation of PR 3.15 at VHDL development platforms, change the values of constant objects "t1" and "max_count" defined in the declarative part of the architecture unit as in

```
        constant t1: natural:=2;
        constant max_count: natural:=2;
```

and the VHDL program in PR 3.15 can be tested using the test-bench in PR 3.16.

```
library ieee;
use ieee.std_logic_1164.all;

entity ss_display_fsm_tb is
end;

architecture bench of ss_display_fsm_tb is

  component ss_display_fsm
    port(clk, rst: in std_logic;
         ssd: out std_logic_vector(6 downto 0));
  end component;

  signal clk, rst: std_logic;
  signal ssd: std_logic_vector(6 downto 0);

  constant clock_period: time:= 500 ms;
  signal stop_the_clock: boolean;

begin
```

```
ps: process   --stimulus
begin
  rst<='1'; rst<='0';
  wait for 9 sec;
  stop_the_clock<=true;
  wait;
end process;

pc: process   --clock generation
begin
  while not stop_the_clock loop
    clk<='0';
    wait for clock_period / 2;
    clk<='1';
    wait for clock_period / 2;
  end loop;
  wait;
end process;

end;
```

PR 3.16 Program 3.16

3.5 The Implementation of Timed Mealy State Machines in VHDL

The implementation of timed Mealy state machines is like the implementation of timed Moore state machines. The only difference occurs in the process unit where circuit outputs and next states are determined. The second process in the template of PR 3.4 can be written for timed Mealy state machines as in PR 3.17.

```
p2: process(present_state, inp1, inp2,....)
begin
 case present_state is
  when st0 =>
   if(inp1=ival1) then
     next_state<=st1;
     timer<=t1;
     outp1<=oval1;  outp2<=oval2; ... outpN<=ovalN;

   elsif(inp1=ival2) then
     next_state<=st2;
     timer<=t2;
     outp1<=oval3;  outp2<=oval4; ... outpN<=ovalM;
         ⋮
   else
     next_state<=stN;
     timer<=tm;
     outp1<=oval5;  outp2<=oval6; ... outpN<=ovalK;
   end if;

  when st1 =>

   if(inp1=ival3) then
     next_state<=st2;
     timer<=t4;
     outp1<=oval7;  outp2<=oval8; ... outpN<=ovalL;
   elsif(inp1=ival4) then
     next_state<=st3;
     timer<=t5;
     outp1<=ova9;  outp2<=oval10; ... outpN<=ovalP;
         ⋮
   else
     next_state<=stN;
     timer<=tm;
     outp1<=oval11;  outp2<=oval12; ... outpN<=ovalR;
   end if;
  when ...
      ⋮
 end case;
end process;
```

PR 3.17   Program 3.17

3.5.1   Example for the VHDL Implementation of Timed Mealy State Machine

In this section, we provide an example for the VHDL implementation of timed Mealy state machine.

Example 3.4 Implement the timed Mealy machine, whose state diagram is given in Fig. 3.14, in VHDL.

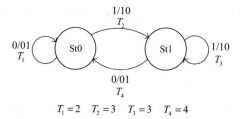

$T_1 = 2 \quad T_2 = 3 \quad T_3 = 3 \quad T_4 = 4$

Fig. 3.14 A timed Mealy state machine/diagram

Solution 3.4 The implementation of the Mealy state machine in Fig. 3.14 is very similar to the implementation of the Moore state machine in Fig. 3.14, the only difference in the implementation arises in the writing of second process, i.e., p2. The second process for Mealy machine is given in PR 3.18.

```
p2: process(present_state, inp)
begin
  case present_state is
    when st0 =>
    if(inp='0') then
      next_state<=st0;
      timer<=t1;
      outp<="01";
    elsif(inp='1') then
      next_state<=st1;
      timer<=t2;
      outp<="10";
    end if;
    when st1 =>
    if(inp='0') then
      next_state<=st0;
      timer<=t4;
      outp<="01";
    elsif(inp='1') then
      next_state<=st1;
      timer<=t3;
      outp<="10";
    end if;
  end case;
end process;
```

PR 3.18 Program 3.18

3.6   Digital Transmitter Implementation Using Timed State Machines

In this section, we explain the VHDL implementation of a digital transmitter using Moore timed state machines.

Example 3.5 The state diagram of a Moore machine is given in Fig. 3.15.

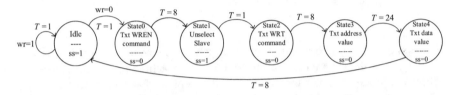

Fig. 3.15 A Moore timed state diagram

The state diagram of Fig. 3.15 belongs to a digital transmitter device. Write-enable command in state0, write command in state2, address in state3, and data in state4 are given as

```
wren="00000000"    wrt="11111111"
addr="1111111110000000011111111"
data="10101010"
```

The bits are transmitted through an output port "outp" at each clock cycle. The entity part of the VHDL implementation of the state diagram of Fig. 3.15 is given in PR 3.19.

```
library ieee;
use ieee.std_logic_1164.all;

entity timed_Moore_fsm is
    port(clk, rst, wr: in std_logic;
         outp, ss: out std_logic );
end entity;
```

PR 3.19 Program 3.19

Complete the VHDL implementation of timed state diagram/machine given in Fig. 3.15.

Solution 3.5 Using the state machine given in Fig. 3.15, we can write the declarative part of the architecture as in PR 3.20.

```
architecture logic_flow of timed_Moore_fsm is

  type state is (idle, st0, st1, st2, st3, st4);
  signal present_state, next_state: state;

  constant t1: natural:=1;
  constant t2: natural:=8;
  constant t3: natural:=24;
  constant max_count: natural:=24;
  signal timer: natural range 0 to max_count;
  signal clk_count: natural range 0 to max_count;

  constant wren: std_logic_vector(7 downto 0):="00000000";
  constant wrt: std_logic_vector(7 downto 0):="11111111";
  constant address: std_logic_vector(23 downto 0):="111111110000000011111111";
  constant data: std_logic_vector(7 downto 0):="10101010";
begin
```

PR 3.20   Program 3.20

The first process, "p1", i.e., state update process is written as in PR 3.21.


```
p1: process(clk, rst)
begin
  if (rst='1') then
    present_state<=idle;
    clk_count<=0;
  elsif (clk'event and clk='1') then
    clk_count<=clk_count+1;
    if(clk_count=timer-1) then
      present_state<=next_state;
      clk_count<=0;
    end if;
  end if;
end process;
```

PR 3.21   Program 3.21

The second process can be written as in PR 3.22.

```
p2: process(present_state, clk_count)          when st2 =>
begin                                            ss<='0'; timer<=t2;
 case present_state is                           next_state<=st3;
                                                 outp<=wrt(7-clk_count);
  when idle =>
   ss<='1'; timer<=t1;                         when st3=>
   if(wr='0') then                               ss<='0'; timer<=t3;
    next_state<=st0;                             next_state<=st4;
   else                                          outp<=address(23-clk_count);
    next_state<=idle;
   end if;                                      when st4 =>
                                                 ss<='0'; timer<=t2;
  when st0 =>                                     next_state<=idle;
   ss<='0'; timer<=t2;                            outp<=data(7-clk_count);
   next_state<=st1;
   outp<=wren(7-clk_count);                     end case;
                                               end process;
  when st1 =>                                 end logic_flow;
   ss<='1'; timer<=t1;
   next_state<=st2;
```

PR 3.22   Program 3.22

Combining all the program parts, we get the complete program as in PR 3.23.

```vhdl
library ieee;
use ieee.std_logic_1164.all;

entity timed_Moore_fsm is
  port(clk, rst, wr: in std_logic;
       outp, ss: out std_logic );
end entity;

architecture logic_flow of timed_Moore_fsm is

  type state is (idle, st0, st1, st2, st3, st4);
  signal present_state, next_state: state;

  constant t1: natural:=1;
  constant t2: natural:=8;
  constant t3: natural:=24;
  constant max_count: natural:=24;
  signal timer: natural range 0 to max_count;
  signal clk_count: natural range 0 to max_count;

  constant wren: std_logic_vector(7 downto 0):="00000000";
  constant wrt: std_logic_vector(7 downto 0):="11111111";
  constant address: std_logic_vector(23 downto 0):="111111110000000011111111";
  constant data: std_logic_vector(7 downto 0):="10101010";
begin
  p1: process(clk, rst)
  begin
   if(rst='1') then
     present_state<=idle;
     clk_count<=0;
   elsif (clk'event and clk='1') then
     clk_count<=clk_count+1;
     if(clk_count=timer-1) then
       present_state<=next_state;
       clk_count<=0;
     end if;
   end if;
  end process;
  p2: process(present_state, clk_count)
  begin
   case present_state is
     when idle =>
       ss<='1'; timer<=t1;
       if(wr='0') then
         next_state<=st0;
       else
         next_state<=idle;
       end if;
     when st0 =>
       ss<='0'; timer<=t2;
       next_state<=st1;
       outp<=wren(7-clk_count);
     when st1 =>
       ss<='1'; timer<=t1;
       next_state<=st2;
     when st2 =>
       ss<='0'; timer<=t2;
       next_state<=st3;
       outp<=wrt(7- clk_count);
     when st3=>
       ss<='0'; timer<=t3;
       next_state<=st4;
       outp<=address(23-clk_count);
     when st4 =>
       ss<='0'; timer<=t2;
       next_state<=idle;
       outp<=data(7-clk_count);

   end case;
  end process;
end logic_flow;
```

PR 3.23   Program 3.23

The VHDL implementation in PR 3.23 can be tested using the test-bench program in PR 3.24.

```vhdl
library ieee;
use ieee.std_logic_1164.all;

entity timed_Moore_fsm_tb is
end;

architecture bench of timed_Moore_fsm_tb is

  component timed_Moore_fsm
    port(clk, rst, wr: in std_logic;
         outp, ss: out std_logic );
  end component;

  signal clk, rst, wr: std_logic;
  signal outp, ss: std_logic;
  constant clock_period: time:= 10 ns;
  signal stop_the_clock: boolean;

begin

  pm: timed_Moore_fsm port map (clk  => clk,
                                rst  => rst,
                                wr   => wr,
                                outp => outp,
                                ss   => ss );
  ps: process   --stimulus
  begin
    rst<='1'; wr<='1';
    rst<='0'; wr<='0';
    wait for clock_period*50;
    stop_the_clock<=true;
    wait;
  end process;

  pc: process   --clock generation
  begin
    while not stop_the_clock loop
      clk<= '0';
      wait for clock_period / 2;
      clk<= '1';
      wait for clock_period / 2;
    end loop;
    wait;
  end process;
end;
```

PR 3.24 Program 3.24

Problems

1. Implement the timed state machine shown in Fig. P3.1 in VHDL, simulate your program using a VHDL development platform such as Vivado, Quartus, or Modelsim, and verify its operation.

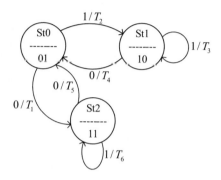

$$T_1 = 5 \quad T_2 = 7 \quad T_3 = 5 \quad T_4 = 6 \quad T_5 = 4 \quad T_6 = 7$$

Fig. P3.1 Moore timed state diagram for P1

2. Convert the Moore state machine shown in Fig. P3.1 to Mealy state machine and implement the Mealy state machine in VHDL.
3. The waveforms in Fig. P3.2 are used between two devices for data transmission. The signals "ss" and "sclk" are used for control purposes, whereas the "data" waveform indicates the transmission of special bit stream. Obtain the timed state machine for the waveforms shown in Fig. P3.2 and implement the timed state machine in VHDL.

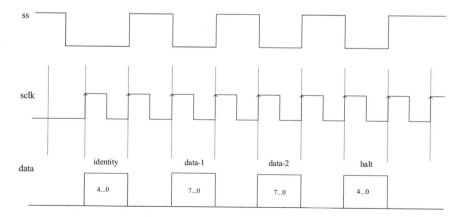

Fig. P3.2 Timing waveform for P3

Serial Peripheral Interface

4

Serial communication can be divided into two main categories, one is the asynchronous communication and the other is the synchronous communication. RS232 communication is an asynchronous communication type in which start and stop bits are used for the controlling of communication, and the use of controlling bits reduces the transmission efficiency. In synchronous serial communication both the transmitter and receiver use the same clock signal. For this reason, in synchronous communication at least two wires, one is used for clock and the other one is used for data, should be available between transmitter and receiver modules. Serial peripheral interface (SPI) is a synchronous communication protocol developed by Motorola company for the synchronous serial communication of 68HC family of microcontrollers by its peripherals. Later on, this standard became de-facto for serial communication, and it has been adapted by many companies for their products. In this chapter, we first give information about synchronous serial communication and SPI protocol, and then explain how to write VHDL codes to implement SPI protocol for FPGA devices so that they can communicate with electronic devices utilizing SPI communication ports.

4.1 Synchronous Communication

In synchronous communication, the transmitter and receiver use the same clock signals for the transmission and reception of information bits. Bits are placed on to the bus at the rising/falling edges of the clock pulses and detected at the falling/rising edges of the clock pulses at the receiver side. For instance, assume that we have the bit sequence

$$d = [d_7\ d_6\ d_5\ d_4\ d_3\ d_2\ d_1\ d_0]$$

© The Author(s), under exclusive license to Springer Nature Switzerland AG 2021
O. Gazi, A. Ç. Arlı, *State Machines using VHDL*,
https://doi.org/10.1007/978-3-030-61698-4_4

and rising edges of the clock pulses are used to transmit data bits. At the rising edge of the first clock pulse, the most significant bit of d is placed onto the bus as indicated in Fig. 4.1.

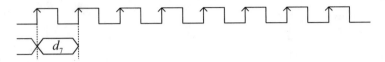

Fig. 4.1 First data bit placed onto the bus

At the rising of the second clock pulse, the information bit d_6 is placed onto the bus as indicated in Fig. 4.2.

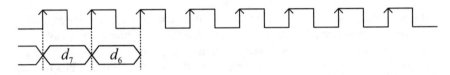

Fig. 4.2 Second data bit placed onto the bus

In a similar manner, the other bits are placed onto the bus at the rising edge of the clock pulses in a sequential manner as shown in Fig. 4.3.

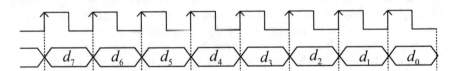

Fig. 4.3 All the data bits placed onto the bus

At the receiver side, the detections of the transmitted bits are performed at the falling edges of the clock pulses. For instance, the detection of d_7 at the receiver side is performed at the falling edge of the first clock pulse as shown in Fig. 4.4.

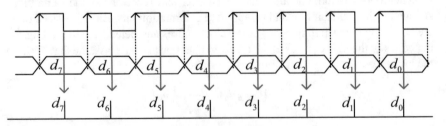

Fig. 4.4 Read operation in synchronous communication

4.2   Serial Peripheral Interface (SPI) Communication

In SPI communication, there is a master which is the source of clock, and a number of slaves which get the clock from master and receive the data transmitted by the master, and they may send data back to master.

An electronic unit may act as only master or as only slave or as both. If an electronic unit can act as a slave only, then the electronic unit does not have the capability of sending its clock to other electronic units. The master unit of the SPI communication is usually a microcontroller and the slaves are the peripherals which can be LCDs, sensors, electronic chips, RFID card readers, wireless transmitters and receivers, and other microcontrollers.

It is possible to transmit data without interruption using any serial synchronous communication protocol like SPI. SPI communication standard employs four-wire bus. In Fig. 4.5, the black-box representation of two electronic units having SPI connectors, and their connection for SPI communication are illustrated.

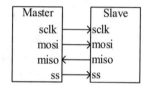

Fig. 4.5 Master-Slave connections

In Fig. 4.5, **sclk** is the serial clock source, **mosi** denotes the master output slave input, **miso** denotes master input slave output, and **ss** denotes slave select. Slave-select output can also be called chip select, i.e., **cs**.

Example 4.1 If the master is a microcontroller and the slave is a simple digital-to-analog (D/A) converter, then the SPI connection between these two devices can be drawn as in Fig. 4.6.

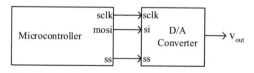

Fig. 4.6 D/A converter as a slave unit

As it is seen from Fig. 4.6 that D/A converter does not have digital output, this means that miso port of the D/A converter is not available.

A master unit can control a number of slaves. In Fig. 4.7, a master is connected to four slaves.

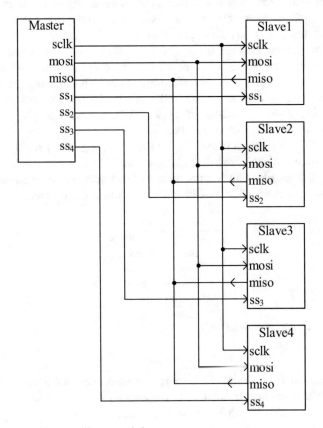

Fig. 4.7   A master unit controlling several slaves

A master with a single slave-select output can be connected to a number of slaves as shown in Fig. 4.8.

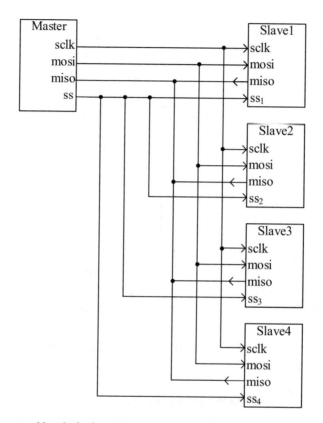

Fig. 4.8 A master with a single slave-select connected to several slaves

Electrically erasable programmable read-only memories (EEPROM) AT25010B/020B/040B, which are organized as 128/256/512 words of 8 bits each, possess SPI ports for serial communication and they can be used as slaves by microcontrollers.

If an electronic unit acts only as a slave, then its black-box representation can be drawn as in Fig. 4.9. The output of one slave can be connected to the input of another slave as in Fig. 4.10.

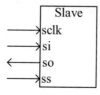

Fig. 4.9 A slave device

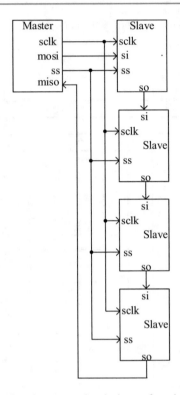

Fig. 4.10 The output of one slave is connected to the input of another slave

4.2.1 MOSI and MISO Bit Transmission

The master unit sends the information stream to the slave bit-by-bit in a serial manner through the MOSI port, and the slave unit receives the data sent from the master using the MOSI port. During the transmission, first, the most significant bit of the data is sent from the master to the slave or from the slave to the master.

4.2.1.1 The Steps of SPI Data Transmission
1. First of all, the synchronization clock should be sent from the master to the slave via **sclk** pin as shown in Fig. 4.11.

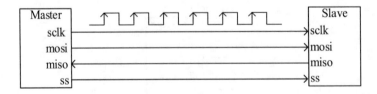

Fig. 4.11 Synchronization clock sent from master to slave

2. In the next step, the master generates the SS pulse which activates the slave as shown in Fig. 4.12. SS signal can be produced using either falling or rising edges of the clock pulses. It is seen in Fig. 4.12 that the SS signal is produced using the falling edges of the clock pulses.

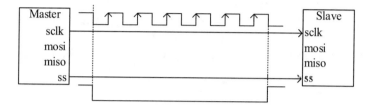

Fig. 4.12 SS pulse sent from master to slave

3. The master transmits the bits using the MOSI port as shown in Fig. 4.13 where the first transmitted bit arrives in the receiver the first.

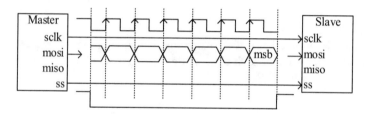

Fig. 4.13 MSB sent from master to slave

4. The slave can also transmit bit sequences to the master as shown in Fig. 4.14.

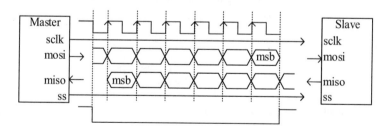

Fig. 4.14 MSB is sent first

4.2.1.2 Properties of SPI Communication

The standard SPI protocol supports up to eight slaves. The maximum distance between a master and a slave can be up to 3 m, or 10 ft approximately, and the maximum speed for SPI communication stayed at 10 Mbps for long time. Certainly, the maximum speed value increases in time.

4.2.2   SPI Operation Modes

Before explaining the SPI operation modes, let us give some information about clock polarity, i.e., CPOL, and clock phase, i.e., CPHA.

4.2.2.1 Clock Polarity (CPOL)

Clock polarity is related to the sensitivity edge of the clock pulse. Pulse sequences with two different clock polarities are shown in Fig. 4.15 where CPOL = 0 indicates the sensitivity to the rising edge, whereas CPOL = 1 indicates the sensitivity to the falling edge.

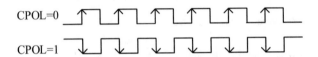

Fig. 4.15 CPOL = 0 and CPOL = 1 illustration

4.2.2.2 Clock Phase

Clock phase is accepted as 0 if the sampling operation at the receiver side is performed at the first edge of each clock pulse. In Fig. 4.16, CPHA = 0 case is illustrated for clock polarity CPOL = 0.

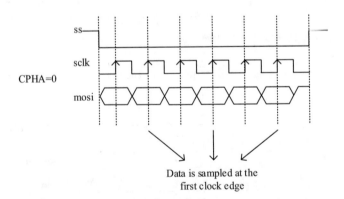

Fig. 4.16 Illustration of Mod-00 SPI operation

On the other hand, if the sampling operation is performed at the second edge of the clock pulse, then the clock phase is accepted as 1. In Fig. 4.17, CPHA = 1 case is illustrated for clock polarity CPL = 0.

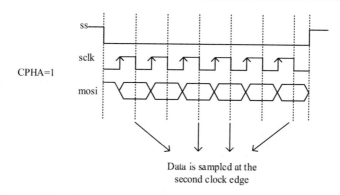

Fig. 4.17 Illustration of Mod-10 SPI operation

The case CPHA = 0, CPOL = 0 is indicated as mode-00 operation of SPI, and similarly for CPHA = 1, CPOL = 0, the transmission operation is indicated as mode-10. When CPOL = 1, we have the transmission schemes shown in Figs. 4.18 and 4.19 for CPHA = 0 and CPHA = 1.

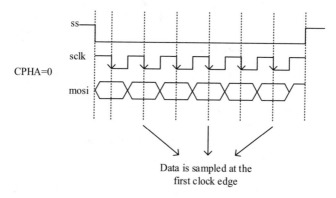

Fig. 4.18 Illustration of Mod-01 SPI operation

The case CPHA = 0, CPOL = 1 is indicated as mode-01 operation of SPI, and similarly for CPHA = 1, CPOL = 1, the transmission operation is indicated as mode-11.

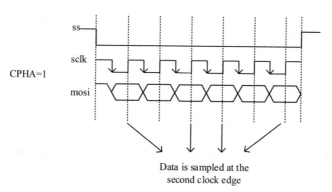

Fig. 4.19 Illustration of Mod-11 SPI operation

The operation modes illustrated in Figs. 4.18 and 4.19 are referred to as mod-01 and mod-11 operation modes.

4.2.2.3 Summary
The operation modes illustrated in Figs. 4.16, 4.17, 4.18, and 4.19 of the SPI serial communications can be summarized as

$$CPHA = 0, CPOL = 0 \Longrightarrow mode\ 00$$
$$CPHA = 0, CPOL = 1 \Longrightarrow mode\ 01$$
$$CPHA = 1, CPOL = 0 \Longrightarrow mode\ 10$$
$$CPHA = 1, CPOL = 1 \Longrightarrow mode\ 11.$$

A master-slave pair must use the same SPI operation mode to communicate. The most used operation modes are mode-00 and mode-11. The modes 00 and 10 are depicted in Fig. 4.20 for comparison purpose.

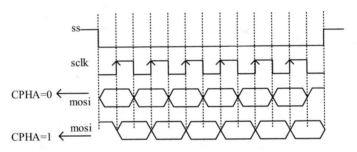

Fig. 4.20 Comparison of Mod-00 and Mod-10 SPI operations

Example 4.2 (a) Draw the waveforms for SPI serial communication for mode-00 operation for the transmission of bit sequence 11010110.

Solution 4.2 First we draw the **ss, sclk** waveforms as shown in Fig. 4.21 where it is seen that **sclk** includes eight pulses, and 8 is the number of data bits to be transmitted.

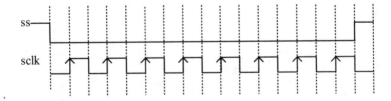

Fig. 4.21 SS and SCLK waveforms

The most significant bit of the sequence $\underline{1}1010110$ is 1. In mod-00 operation, the data bit is read at the first edge of each clock pulse. This means that data bit is placed onto the bus at the falling edge, and it is read at the receiver at the rising edge of the clock pulse. Considering this, the transmission of the MSB, i.e., bit-7, is depicted in Fig. 4.22.

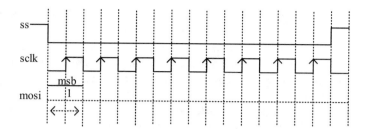

Fig. 4.22 MSB transmission

The transmission of the bit-6 is illustrated in Fig. 4.23.

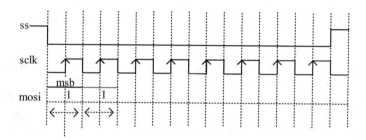

Fig. 4.23 The transmission of the bit-6

Following a similar logic for the rest of the bits, the **mosi** waveform which depicts the transmission of the sequence 11010110 together with the **ss** and **sclk** waveforms can be drawn as in Fig. 4.24.

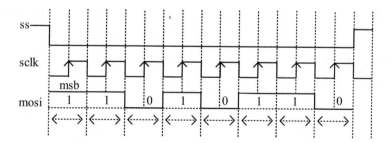

Fig. 4.24 The transmission of all the data bits

At the backside of the SPI port, we have several registers, and these registers are used to determine the communication parameters of the SPI communication. These parameters are speed in bits/s, mode of operation, values of control, and status bits. Usually in SPI communication, before the exchange of data between master and slave, master sends some known bit sequences to the slave to program its registers for SPI communication.

The most common registers are data, control, and status registers, and these registers are not fixed in number in every SPI device. They may vary in number and type according to the manufacturers of the device. For instance, Maxim DS1306 RTC has control and status registers, and FM25L512 FRAM memory owns only a single status register.

4.3   VHDL Implementation of SPI Communication

In this section, we will explain the VHDL implementation of the SPI communication protocol. SPI communication can be implemented using state or timed state machines. For simple applications, classical state machines can be utilized. In general, timed state machines are used for the VHDL implementation of SPI communication. We will explain the subject via examples. First, we will consider the implementation of SPI protocols which only transmit data, then we will implement the SPI protocols which both transmit and receive data.

4.3.1   Implementation of SPI Protocols Only in Transmit Mode

In this section, we will implement SPI protocol in VHDL considering only master-to-slave data transmission scenario. We will explain the subject through an example.

Example 4.3 Draw the state machine diagram for the transmission of the bit sequence 11010110 using the SPI protocol in mod-00 operation type.

Solution 4.3 SPI waveforms are given in Fig. 4.24. Considering the waveforms in Fig. 4.24, we can draw the Moore state diagram as in Fig. 4.25.

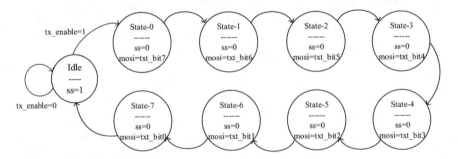

Fig. 4.25 A Moore state machine

Example 4.4 Implement the transmission of bit sequence 11010110 using the SPI protocol in mod-00 operation in VHDL. Assume that the clock frequency of SP equals to the clock frequency of FPGA.

Solution 4.4 SPI protocol can be implemented using state machines. The state diagram for the transmission of bit sequence 11010110 using the SPI protocol is depicted in Fig. 4.25. Considering Fig. 4.25, first, we write the entity part and declarative part of the architecture as in PR 4.1.

```
library ieee;
use ieee.std_logic_1164.all;
entity fsm_spi is
  port(clk, rst, tx_enable: in std_logic;
       mosi, ss, sclk: out std_logic );
end entity;

architecture logic_flow of fsm_spi is
  type state is (st_idle, st_txBit7, st_txBit6, st_txBit5, st_txBit4,
                 st_txBit3, st_txBit2, st_txBit1, st_txBit0);
  signal present_state, next_state: state;
  constant data: std_logic_vector(7 downto 0) :="11010110";
  signal spi_sclk: std_logic;
begin
```

PR 4.1 Program 4.1

In PR 4.1, "spi_sclk" is the signal object used for the clock of the SPI protocol, and for simplicity we assume that "frequency" of "spi_sclk" equals to the clock frequency of FPGA. "sclk" is the clock signal sent to the receiver. If "spi_sclk" is not equal to the clock frequency of FPGA, we need to use a frequency divider to obtain "spi_sclk" from FPGA's clock frequency "clk".

Our complete program includes two processes, one of them is for the update of the present state, and the other is for the calculation of the next states and circuit outputs. The structure of the overall program happens as in PR 4.2.

```
library ieee;
use ieee.std_logic_1164.all;
entity fsm_spi is
  port( clk, rst, tx_enable: in std_logic;
        mosi, ss, sclk: out std_logic );
end entity;

architecture logic_flow of fsm_spi is
  type state is (st_idle, st_txBit7, st_txBit6, st_txBit5,
                 st_txBit4, st_txBit3, st_txBit2, st_txBit1, st_txBit0);
  signal present_state, next_state: state;
  constant data: std_logic_vector(7 downto 0):="11010110";
  signal spi_sclk: std_logic;
begin
  spi_sclk<=clk;
  sclk<=spi_sclk;

p1: process(spi_sclk, rst)
--- present state update
begin
  .
  .
end process;

p2: process(present_state, tx_enable)
--- outputs and next states
begin
  .
  .
end process;

end architecture;
```

PR 4.2 Program 4.2

The update of the present state is performed in PR 4.3.

```
p1: process(spi_sclk, rst)  --present state update
begin
  if(rst='1' or (tx_enable ='0')) then
    present_state<=st_idle;
  elsif(spi_sclk'event and spi_sclk='0') then
    present_state<=next_state;
  end if;
end process;
```

PR 4.3   Program 4.3

SPI transmission protocol is implemented in VHDL in PR 4.4.

```
--- Circuit outputs and next states
p2: process(present_state, tx_enable)
begin
  case present_state is
    when st_idle =>
      ss<='1'; sclk<='0'; mosi<='X';
      if(tx_enable ='1') then
        next_state<=st_txBit7;
      else
        next_state<=st_idle;
      end if;

    when st_txBit7=>
      ss<='0';
      mosi<=data(7);
      next_state<=st_txBit6;

    when st_txBit6=>
      ss<='0';
      mosi<=data(6);
      next_state<=st_txBit5;

    when st_txBit5=>
      ss<='0';
      mosi<=data(5);
      next_state<=st_txBit4;

    when st_txBit4=>
      ss<='0';
      mosi<=data(4);
      next_state<=st_txBit3;

    when st_txBit3=>
      ss<='0';
      mosi<=data(3);
      next_state<=st_txBit2;

    when st_txBit2=>
      ss<='0';
      mosi<=data(2);
      next_state<=st_txBit1;

    when st_txBit1=>
      ss<='0';
      mosi<=data(1);
      next_state<=st_txBit0;

    when st_txBit0=>
      ss<='0';
      mosi<=data(0);
      next_state<=st_idle;
  end case;
end process;
```

PR 4.4   Program 4.4

The complete program happens to be as in PR 4.5.

```vhdl
library ieee;
use ieee.std_logic_1164.all;
entity fsm_spi is
  port(clk, rst, tx_enable:  in std_logic;
        mosi, ss, sclk: out  std_logic );
end entity;

architecture logic_flow of fsm_spi is
  type state is (st_idle, st_txBit7, st_txBit6, st_txBit5,
                 st_txBit4, st_txBit3,st_txBit2,st_txBit1,st_txBit0);
  signal present_state, next_state: state;

  constant data: std_logic_vector(7 downto 0):="11010110";
  signal spi_sclk: std_logic;

begin

  spi_sclk<=clk;
  sclk<=spi_sclk;

  p1: process(spi_sclk, rst) -- The next state logic
  begin
    if(rst='1' or (tx_enable ='0')) then
      present_state<= st_idle;
    elsif (spi_sclk'event and spi_sclk='0') then
      present_state<=next_state;
    end if;
  end process;

--- Circuit outputs and next state values
  p2: process(present_state, tx_enable)
  begin
    case present_state is
      when st_idle =>
        ss<='1';  sclk<='0';  mosi<='X';
        if(tx_enable ='1') then
          next_state<= st_txBit7;
        else
          next_state<= st_idle;
        end if;

      when st_txBit7=>
        ss<='0';
        mosi<= data(7);
        next_state<= st_txBit6;

      when st_txBit6=>
        ss<='0';
        mosi<= data(6);
        next_state<= st_txBit5;

      when st_txBit5=>
        ss<='0';
        mosi<= data(5);
        next_state<= st_txBit4;

      when st_txBit4=>
        ss<='0';
        mosi<= data(4);
        next_state<= st_txBit3;

      when st_txBit3=>
        ss<='0';
        mosi<= data(3);
        next_state<= st_txBit2;

      when st_txBit2=>
        ss<='0';
        mosi<= data(2);
        next_state<= st_txBit1;

      when st_txBit1=>
        ss<='0';
        mosi<= data(1);
        next_state<= st_txBit0;

      when st_txBit0=>
        ss<='0';
        mosi<= data(0);
        next_state<= st_idle;
    end case;
  end process;
end logic_flow;
```

PR 4.5 Program 4.5

The program of PR 4.5 can be tested using the test-bench in PR 4.6.

```
library ieee;
use ieee.std_logic_1164.all;

entity spi_Example_TB is
end spi_Example_TB;

architecture logic_flow of spi_Example_TB is

  component fsm_spi
   port(clk, rst, tx_enable: in std_logic;
        mosi, ss, sclk: out std_logic );
  end component;

  signal clk1, rst1, tx_enable1: std_logic:='0';
  signal mosi1, ss1, sclk1: std_logic:='0';


  constant clock_period: time :=10 ns;
  signal stop_the_clock: boolean;


begin

  pm: fsm_spi port map(clk=>clk1,
                       rst=>rst1,
                       tx_enable=>tx_enable1,
                       mosi=>mosi1,
                       ss=>ss1,
                       sclk=>sclk1);
```

```
ps: process
begin

  rst1<='1';
  wait for clock_period;

  rst1<='0';
  tx_enable1<='1';
  wait for clock_period;

  wait for clock_period*8;

  stop_the_clock<=true;
  wait;

end process;

pc: process  --clock generations
begin
  while not stop_the_clock loop
    clk1<='0';
    wait for clock_period / 2;
    clk1<='1';
    wait for clock_period / 2;
  end loop;
  wait,
end process;

end architecture;
```

PR 4.6 Program 4.6

Instead of using separate two processes, we can also use a single process for the implementation of state machines. We can combine the processes in PR 4.3 and PR 4.4 as in PR 4.7 where it is seen that the sensitivity list of the merged process contains more parameters.

```vhdl
library ieee;
use ieee.std_logic_1164.all;
entity fsm_spi is
  port(clk, rst, tx_enable:  in std_logic;
       mosi, ss, sclk: out std_logic );
end entity;

architecture logic_flow of fsm_spi is
  type state is(st_idle, st_txBit7, st_txBit6,st_txBit5,
               st_txBit4,st_txBit3,st_txBit2,st_txBit1,st_txBit0);
  signal present_state, next_state: state;

  constant data: std_logic_vector(7 downto 0):="11010110";
  signal spi_sclk: std_logic;

begin

  spi_sclk<=clk;
  sclk<=spi_sclk;

  p1_p2: process(spi_sclk, rst, present_state, tx_enable)
  begin

    if(rst='1' or (tx_enable ='0')) then
      present_state<=st_idle;
    elsif (spi_sclk'event and spi_sclk='0') then
      present_state<=next_state;
    end if;

    case present_state is
      when st_idle =>
        ss<='1'; mosi<='X';
        if(tx_enable ='1') then
          next_state<=st_txBit7;
        else
          next_state<=st_idle;
        end if;

      when st_txBit7=>
        ss<='0';
        mosi<=data(7);
        next_state<=st_txBit6;
```

```vhdl
      when st_txBit6=>
        ss<='0';
        mosi<=data(6);
        next_state<=st_txBit5;

      when st_txBit5=>
        ss<='0';
        mosi<=data(5);
        next_state<=st_txBit4;

      when st_txBit4=>
        ss<='0';
        mosi<=data(4);
        next_state<=st_txBit3;

      when st_txBit3=>
        ss<='0';
        mosi<=data(3);
        next_state<=st_txBit2;

      when st_txBit2=>
        ss<='0';
        mosi<=data(2);
        next_state<=st_txBit1;

      when st_txBit1=>
        ss<='0';
        mosi<=data(1);
        next_state<=st_txBit0;

      when st_txBit0=>
        ss<='0';
        mosi<=data(0);
        next_state<=st_idle;
    end case;
  end process;
end logic_flow;
```

PR 4.7 Program 4.7

4.3.1.1 Second Solution

In the previous solution, too many states are used in the program unit. To decrease the number of program lines, we can consider the use of timed state machines for the implementation of the SPI protocol for the given bit stream. The state diagram of the SPI protocol for the transmission of the bit stream 11010110 can be drawn as in Fig. 4.26.

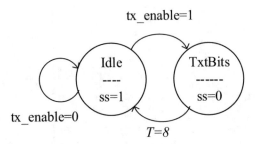

Fig. 4.26 State diagram for second solution

Using the state diagram of Fig. 4.26, we can implement the SPI protocol for the transmission of bit stream 11010110 as follows. First, we write the entity unit and declarative part of the architecture as in PR 4.8. It is assumed that clock frequency of the SPI protocol equals to the clock frequency of the FPGA device. Otherwise, we need to use a clock divider which generates SPI's clock frequency from FPGA's clock frequency.

```
library ieee;
use ieee.std_logic_1164.all;
entity fsm_spi is
  port(clk, rst, tx_enable: in std_logic;
        mosi, ss, sclk: out  std_logic );
end entity;

architecture logic_flow of  fsm_spi is
  type state is (st_idle, st_txmt);
  signal present_state, next_state: state;

  constant data: std_logic_vector(7 downto 0):="11010110";

  constant data_length: natural:=8;
  signal timer: natural range 0 to data_length;
  signal data_index: natural range 0 to data_length;
  signal spi_sclk: std_logic;
begin
  spi_sclk<=clk;
  sclk<=spi_sclk;
```

PR 4.8 Program 4.8

The next state logic of the program can be written as in PR 4.9.

```
p1: process(spi_sclk, rst)
begin
  if (rst='1') then
    present_state<=st_idle;
    data_index<= 0;
  elsif (spi_sclk'event and spi_sclk='0') then
    if(data_index=timer-1) then
      present_state<=next_state;
      data_index<=0;
    else
      data_index<=data_index +1;
    end if;
  end if;
end process;
```

PR 4.9 Program 4.9

The outputs of the SPI protocol using the state machines can be implemented as in PR 4.10.

```
--- Circuit outputs and next states
p2: process(present_state, tx_enable, data_index)
begin
  case present_state is
    when st_idle =>
      ss<='1';
      mosi<='X';
      timer<=1;
      if(tx_enable='1') then
        next_state<=st_txmt;
      else
        next_state<=st_idle;
      end if;

    when st_txmt =>
      ss<='0';
      timer<=8;
      mosi<=data(7-data_index);
      next_state<=st_idle;

  end case;
end process;
end logic_flow;
```

PR 4.10 Program 4.10

The complete program happens to be as in PR 4.11.

```vhdl
library ieee;
use ieee.std_logic_1164.all;

entity fsm_spi is
  port(clk, rst, tx_enable:  in std_logic;
       mosi, ss, sclk: out  std_logic );
end entity;

architecture logic_flow of fsm_spi is
  type state is (st_idle, st_txmt);
  signal present_state, next_state: state;
  constant data: std_logic_vector(7 downto 0):="11010110";
  constant data_length: natural:=8;
  signal timer: natural range 0 to data_length;
  signal data_index: natural range 0 to data_length;
  signal spi_sclk: std_logic;
begin
  spi_sclk<=clk;  sclk<=spi_sclk;
  p1: process(spi_sclk, rst)
  begin
   if(rst='1') then
     present_state<=st_idle;
     data_index<= 0;
    elsif(spi_sclk'event and spi_sclk='0') then
     if(data_index=timer-1) then
       present_state<=next_state;
       data_index<=0;
     else
       data_index<=data_index +1;
     end if;
   end if;
  end process;
  p2: process(present_state, tx_enable, data_index)
  begin
   case present_state is
     when st_idle =>
     ss<='1';  mosi<='X';
     timer<=1;
     if(tx_enable='1') then
       next_state<=st_txmt;
     else
       next_state<=st_idle;
     end if;
     when st_txmt =>
     ss<='0'; timer<=8;
     mosi<=data(7-data_index);
     next_state<=st_idle;
   end case;
  end process; end;
```

PR 4.11 Program 4.11

The VHDL implementation in PR 4.11 can be tested using the test-bench program given in PR 4.12.

```vhdl
library ieee;
use ieee.std_logic_1164.all;

entity fsm_spi_tb is
end;

architecture bench of fsm_spi_tb is

  component fsm_spi
    port(clk, rst, tx_enable: in std_logic;
         mosi, ss, sclk: out std_logic );
  end component;

  signal clk, rst, tx_enable: std_logic:='0';
  signal mosi, ss, sclk: std_logic:='0';
  constant clock_period: time:= 10 ns;
  signal stop_the_clock: boolean;

begin
  pm: fsm_spi port map(clk       => clk,
                       rst       => rst,
                       tx_enable => tx_enable,
                       mosi      => mosi,
                       ss        => ss,
                       sclk      => sclk);




ps: process --stimulus
begin
  rst<='1';
  wait for clock_period;

  rst<='0';
  tx_enable<='1';
  wait for clock_period;

  wait for clock_period*8;

  tx_enable<='0';
  stop_the_clock<=true;
  wait;
end process;

pc: process --clock generation
begin
  while not stop_the_clock loop
    clk<='0';
    wait for clock_period / 2;
    clk<='1';
    wait for clock_period / 2;
  end loop;
  wait;
end process;
end;
```

PR 4.12 Program 4.12

Instead of two processes, we can also use a single process for the implementation of state machine. We can combine the processes in PR 4.12 in PR 4.13.

```
p1_p2: process(spi_sclk, present_state, rst, tx_enable, data_index)
begin

 if(rst='1') then
   present_state<=st_idle;
   data_index<= 0;
 elsif (spi_sclk'event and spi_sclk='0') then
   if(data_index=timer-1) then
     present_state<=next_state;
     data_index<=0;
   else
     data_index<=data_index +1;
   end if;
 end if;

 case present_state is
   when st_idle =>
   ss<='1';
   mosi<='X';
   timer<=1;
   if(tx_enable='1') then
     next_state<=st_txmt;
   else
     next_state<=st_idle;
   end if;

   when st_txmt =>
   ss<='0';
   timer<=8;
   mosi<=data(7- data_index);
   next_state<=st_idle;
 end case;
end process;
```

PR 4.13   Program 4.13

Example 4.5 Assume that a master unit, connected to a slave device, sends a 4-bit
control sequence followed by an 8-bit data sequence using SPI communication protocol. Implement the SPI transmission protocol in VHDL. Use mod-00.

Solution 4.5 The timing waveforms for the SPI protocol can be drawn as in
Fig. 4.27.

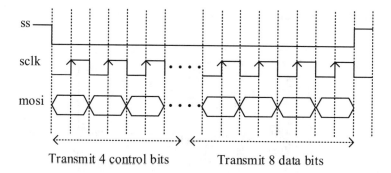

Fig. 4.27 The timing waveforms for the SPI protocol for Example 4.5

The state diagram of the SPI transmission protocol can be drawn as in Fig. 4.28.

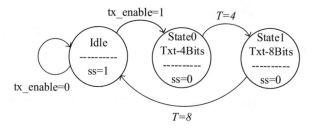

Fig. 4.28 The state diagram of SPI protocol for Example 4.5

First, we write the entity unit and declarative part of the architecture as in PR 4.14.

```
library ieee;
use ieee.std_logic_1164.all;
entity fsm_spi is
  port( clk, rst, tx_enable: in std_logic;
        mosi, ss, sclk: out std_logic );
end entity;

architecture logic_flow of fsm_spi is
  type state is (st_idle, st0_txmt, st1_txmt);
  signal present_state, next_state: state;

  constant control: std_logic_vector(3 downto 0):="1110";
  constant data: std_logic_vector(7 downto 0):="11010110";

  constant max_length: natural:=8;
  signal timer: natural range 0 to max_length;
  signal data_index: natural range 0 to max_length;

  signal spi_sclk: std_logic;
begin
```

PR 4.14 Program 4.14

The next state logic of the program can be written as in PR 4.15.

```
p1: process(spi_sclk, rst)
begin
  if(rst='1') then
    present_state<=st_idle;
    data_index<=0;
  elsif (spi_sclk'event and spi_sclk='0') then
    if(data_index=timer-1) then
      present_state<=next_state;
      data_index<=0;
    else
      data_index<= data_index +1;
    end if;
  end if;
end process;
```

PR 4.15 Program 4.15

The outputs of the SPI protocol using the state machines can be implemented as in PR 4.16.

```
--- Circuit outputs and next states
p2: process(present_state, tx_enable, data_index)
begin
  case present_state is
    when st_idle =>
      ss<='1';
      mosi<='X';
      timer<=1;
      if(tx_enable ='1') then
        next_state<=st0_txmt;
      else
        next_state<=st_idle;
      end if;

    when st0_txmt =>
      ss<='0';
      timer<=4;
      mosi<=control(3- data_index);
      next_state<=st1_txmt;

    when st1_txmt =>
      ss<='0';
      timer<=8;
      mosi<=data(7- data_index);
      next_state<=st_idle;
  end case;
end process;
end logic_flow;
```

PR 4.16 Program 4.16

Combining all the program parts, we get the overall program as in PR 4.17.

```vhdl
library ieee;
use ieee.std_logic_1164.all;
entity fsm_spi is
  port(clk, rst, tx_enable: in std_logic;
      mosi, ss, sclk: out std_logic );
end entity;


architecture logic_flow of fsm_spi is
  type state is (st_idle, st0_txmt, st1_txmt);
  signal present_state, next_state: state;

  constant control: std_logic_vector(3 downto 0):="1110";
  constant data: std_logic_vector(7 downto 0):="11010110";

  constant max_length: natural:=8;
  signal timer: natural range 0 to max_length;
  signal data_index: natural range 0 to max_length;

  signal spi_sclk: std_logic;
begin
spi_sclk<=clk;
sclk<=spi_sclk;

p1: process(spi_sclk, rst)
begin
  if(rst='1') then
    present_state<=st_idle;
    data_index<=0;
  elsif(spi_sclk'event and spi_sclk='0') then
    if(data_index=timer-1) then
      present_state<=next_state;
      data_index<=0;
    else
      data_index<=data_index +1;
    end if;
  end if;
end process;

p2: process(present_state, tx_enable,
                data_index)
begin
  case present_state is
    when st_idle =>
      ss<='1';
      mosi<='X';
      timer<=1;
      if(tx_enable ='1') then
        next_state<=st0_txmt;
      else
        next_state<= st_idle;
      end if;
    when st0_txmt =>
      ss<='0';
      timer<=4;
      mosi<=control(3-data_index);
      next_state<=st1_txmt;

    when st1_txmt =>
      ss<='0';
      timer<=8;
      mosi<=data(7-data_index);
      next_state<=st_idle;

  end case;
end process;

end logic_flow;
```

PR 4.17 Program 4.17

The VHDL implementation in PR 4.17 can be tested using the test-bench in PR 4.18.

```vhdl
library ieee;
use ieee.std_logic_1164.all;

entity fsm_spi_tb is
end;

architecture bench of fsm_spi_tb is
  component fsm_spi
    port(clk, rst, tx_enable:  in std_logic;
         mosi, ss, sclk: out  std_logic );
  end component;

  signal clk, rst, tx_enable: std_logic;
  signal mosi, ss, sclk: std_logic ;
  constant clock_period: time:= 10 ns;
  signal stop_the_clock: boolean;

begin
  pm: fsm_spi port map(clk      => clk,
                       rst      => rst,
                       tx_enable=> tx_enable,
                       mosi     => mosi,
                       ss       => ss,
                       sclk     => sclk);
  ps: process   -- stimulus
  begin
    rst<='1';
    wait for clock_period;
    rst<='0';
    tx_enable<='1';
    wait for clock_period;
    wait for clock_period*12;
    tx_enable<='0';
    stop_the_clock<=true;
    wait;
  end process;

  pc: process -- clock generation
  begin
    while not stop_the_clock loop
      clk<='0';
      wait for clock_period / 2;
      clk<='1';
      wait for clock_period / 2;
    end loop;
    wait;
  end process;
end;
```

PR 4.18 Program 4.18

In PR 4.17, instead of using two processes, we can also use single process as in PR 4.19 where it is seen that the body part of two processes are combined, and their sensitivity lists are merged.

```
p1_p2:  process(spi_sclk, rst, present_state,
                tx_enable, data_index)

-- Circuit outputs and next state values
begin

 if(rst='1') then
  present_state<=st_idle;
  data_index<= 0;
 elsif (spi_sclk'event and spi_sclk='0') then
  if(data_index=timer-1) then
   present_state<=next_state;
   data_index <=0;
  else
   data_index<=data_index +1;
  end if;
 end if;

 case present_state is
  when st_idle =>
   ss<='1';
   mosi<='X';
   timer<=1;
```

```
  if(tx_enable='1') then
   next_state<=st0_txmt;
  else
   next_state<=st_idle;
  end if;

  when st0_txmt =>
   ss<='0';
   timer<=4;
   mosi<=control(3-data_index);
   next_state<=st1_txmt;

  when st1_txmt =>
   ss<='0';
   timer<=8;
   mosi<=data(7-data_index);
   next_state<=st_idle;

 end case;
end process;
```

PR 4.19 Program 4.19

4.3.2 Implementation of SPI Protocols Both in Transmit and Receive Mode

Up to now, we considered the VHDL implementation of the SPI protocols dealing with only data transmission from master to slave. In this section, we will consider the implementation of SPI protocols which both transmit and receive data. We will explain the subject through examples.

Example 4.6 Assume that you have a serial flash memory having SPI communication utility. The flash memory can be read and written. The read operation can be performed using the timing waveforms shown in Fig. 4.29 where "rd_enable" is the control signal for read operation. First READ command is sent for the read operation, and this is followed by a 24 bits of address information, and the incoming bits are read byte-by-byte till the SS signal is asserted high.

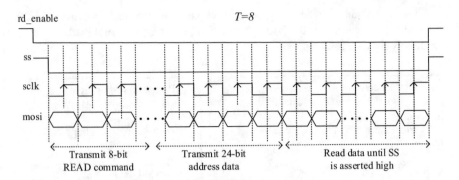

Fig. 4.29 SPI protocol waveforms involving transmit and receive operations

Implement the described SPI protocol in VHDL. Use any values for the 8-bit READ command and 24-bits address value and read a single 8-bit data. Assume that FPGA's clock frequency is 100 MHz, and SPI serial clock is 1 MHz.

Solution 4.6 We can achieve the VHDL implementation of the time waveforms using the timed state machines. The diagram of the timed state machine for the read operation can be drawn as in Fig. 4.30.

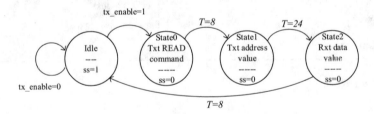

Fig. 4.30 SPI state diagram involving transmit and receive operations

The serial clock of the SPI can be generated from the clock of FPGA using a frequency divider.

The entity part of the timed state machine can be written as in PR 4.20.

```
library ieee;
use ieee.std_logic_1164.all;
entity fsm_spi is
  port(clk_100MHz, rst, tx_enable, miso:  in std_logic;
       mosi, ss, sclk_1MHz: out  std_logic;
       data_read: out std_logic_vector(7 downto 0)  );
end entity;

architecture logic_flow of fsm_spi is
  type state is (st_idle, st0_txRead, st1_txAddress, st2_rxData);
  signal present_state, next_state: state;

  constant read_cmd: std_logic_vector(7 downto 0):="11101100";
  constant address: std_logic_vector(23 downto 0):="110101101101011011010110";

  constant max_length: natural:=24;
  signal data_index: natural range 0 to max_length;
  signal timer: natural range 0 to max_length;

  signal count: natural range 1 to 50;
  signal spi_sclk: std_logic;

begin
```

PR 4.20   Program 4.20

The serial clock of SPI can be obtained using the clock divider process in PR 4.21.

```
p_cdiv: process(clk_100MHz)
begin
  if(rising_edge(clk_100MHz)) then
   count<=count + 1;
   if(count=50) then
    spi_sclk<=not spi_sclk;
    count<=1;
   end if;
  end if;
end process;
sclk_1MHz<=spi_sclk;
```

PR 4.21   Program 4.21

The next state logic of the program can be written as in PR 4.22.

```
p1: process(spi_sclk, rst)
begin
 if(rst='1') then
   present_state<=st_idle;
   data_index<=0;
  elsif(spi_sclk 'event and spi_sclk='0') then
   if(data_index=timer-1) then
    present_state<=next_state;
    data_index<=0;
   else
    data_index<=data_index +1;
   end if;
  end if;
end process;
```

PR 4.22 Program 4.22

Data sending and data receiving operations were implemented in PR 4.23 with the processes "p2" and "p3".

```
--- Circuit outputs and next state values
p2: process(present_state, tx_enable, data_index)
begin
 case present_state is
  when st_idle =>
   ss<='1';
   mosi<='X';
   timer<=1;
   if(tx_enable ='1') then
    next_state<=st0_txRead;
   else
    next_state<=st_idle;
   end if;

  when st0_txRead =>
   ss<='0';
   timer<=8;
   mosi<=read_cmd(7- data_index);
   next_state<=st1_txAddress;

  when st1_txAddress =>
   ss<='0';
   timer<=24;
```

```
   mosi<=address(23-data_index);
   next_state<=st2_rxData;

  when st2_rxData =>
   ss<='0';
   timer<=8;
   next_state<=st_idle;
  end case;

end process;

p3: process(spi_sclk)
begin
 if(spi_sclk'event and spi_sclk='1') then
  if(present_state=st2_rxData) then
   data_read(7-data_index)<=miso;
  end if;
 end if;
end process;
```

PR 4.23 Program 4.23

Combining all the parts, we get the overall program in PR 4.24.

```vhdl
library ieee;
use ieee.std_logic_1164.all;
entity fsm_spi is
  port(clk_100MHz, rst, tx_enable, miso:  in std_logic;
       mosi, ss, sclk_1MHz: out  std_logic; data_read: out std_logic_vector(7 downto 0));
end entity;

architecture logic_flow of fsm_spi is
  type state is (st_idle, st0_txRead, st1_txAddress, st2_rxData);
  signal present_state, next_state: state;

  constant read_cmd: std_logic_vector(7 downto 0):="11101100";
  constant address: std_logic_vector(23 downto 0):="110101101101011011010110";

  constant max_length: natural:=24;
  signal data_index, timer: natural range 0 to max_length;
  signal count: natural range 1 to 50;
  signal spi_sclk: std_logic:='0';
begin
```

```vhdl
p_cdiv: process(rst, clk_100MHz)
begin
 if(rst='1') then
  count<=1;
 end if;
  if(rising_edge(clk_100MHz)) then
   count<=count + 1;
   if(count= 50) then
    spi_sclk<=not spi_sclk;  count<=1;
   end if;
  end if;
 end process;
sclk_1MHz<= spi_sclk;
p1: process(spi_sclk, rst)
begin
 if(rst='1') then
  present_state<=st_idle; data_index<= 0;
 elsif(spi_sclk'event and spi_sclk='0') then
  if(data_index=timer-1) then
   present_state<=next_state; data_index<=0;
  else
   data_index<=data_index +1;
  end if;
 end if;
end process;
p3: process(spi_sclk)
begin
 if(spi_sclk'event and spi_sclk='1') then
  if(present_state=st2_rxData) then
   data_read(7-data_index)<=miso;
  end if;
 end if;
end process;
```

```vhdl
p2: process(present_state, tx_enable, data_index)
begin
  case present_state is
   when st_idle =>
    ss<='1';
    mosi<='X';
    timer<=1;
    if(tx_enable='1') then
     next_state<=st0_txRead;
    else
     next_state<=st_idle;
    end if;
   when st0_txRead =>
    ss<='0';
    timer<=8;
    mosi<=read_cmd(7-data_index);
    next_state<=st1_txAddress;
   when st1_txAddress =>
    ss<='0';
    timer<=24;
    mosi<=address(23-data_index);
    next_state<=st2_rxData;
   when st2_rxData =>
    ss<='0';
    timer<=8;
    next_state<=st_idle;
  end case;
 end process;
end logic_flow;
```

PR 4.24   Program 4.24

The VHDL implementation in PR 4.24 can be tested using the test-bench program given in PR 4.25.

```vhdl
library ieee;
use ieee.std_logic_1164.all;

entity fsm_spi_tb is
end;

architecture bench of fsm_spi_tb is

  component fsm_spi
    port(clk_100MHz, rst, tx_enable, miso: in std_logic;
         mosi, ss, sclk_1MHz: out std_logic;
         data_read: out std_logic_vector(7 downto 0));
  end component;

  signal clk_100MHz, rst, tx_enable, miso: std_logic:='0';
  signal mosi, ss, sclk_1MHz: std_logic:='0';
  signal data_read: std_logic_vector(7 downto 0);

  constant clock_period: time := 10 ns; -- 100MHz clock frequency
  signal stop_the_clock: boolean;

begin
  pm: fsm_spi port map (clk_100MHz => clk_100MHz,
                        rst        => rst,
                        tx_enable  => tx_enable,
                        miso       => miso,
                        mosi       => mosi,
                        ss         => ss,
                        sclk_1MHz  => sclk_1MHz,
                        data_read  => data_read );
  ps: process    -- stimulus:
  begin

  rst<='1';  tx_enable<='0';
  wait for 1 us;-- wait for clock_period*100;

  rst<='0';  tx_enable<='1';
  wait for 1 us;
  wait for 8 us;  wait for 24 us;
```

PR 4.25 Program 4.25

```
miso<='1'; wait for 1 us;  miso<='0'; wait for 1 us;
miso<='1'; wait for 1 us;  miso<='0'; wait for 1 us;
miso<='1'; wait for 1 us;  miso<='0'; wait for 1 us;
miso<='1'; wait for 1 us;  miso<='0'; wait for 1 us;

wait for 16 us;
stop_the_clock<=true;
wait;
end process;

pc: process       clocking
begin
  while not stop_the_clock loop
    clk_100MHz <= '0';
    wait for clock_period / 2;
    clk_100MHz <= '1';
    wait for clock_period / 2;
  end loop;
wait;
end process;
end;
```

PR.4.25 (continued)

4.4 SPI VHDL Implementation Examples for Electronic Devices

In this section, we will provide examples for VHDL implementation of SPI protocol to communicate with some electronic devices commercially available.

4.4.1 VHDL Implementation of SPI Protocol for 12-bit DAC MCP4921

The first devices we consider are MCP4921/4922 which are 12-bit DACs with SPI interface. MCP4922 contains two channel outputs. The black-box representation of MCP4921 is shown in Fig. 4.31.

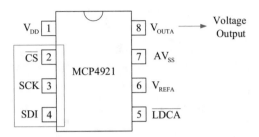

Fig. 4.31 The black-box representation of MCP4921

MCP4921 is a digital-to-analog converter (DAC) which get 12 bits of digital input and produces an analog voltage proportional to the magnitude of the digital input. SPI interface of the MCP4921 consists of chip select $\overline{CS}$, i.e., SS, serial clock SCK, i.e., SCLK, serial data input SDI, i.e., MOSI.

The device does not have MISO output port, i.e., it does not have digital output. It only accepts digital input and produces analog output. V_{DD} is supply voltage which takes values from 2.7 to 5.5 V. V_{OUTA} is the output voltage. V_{REFA} is the voltage reference input which ranges from AV_{SS}, the analog ground input, to V_{DD}.

The MCP4921 supports mode-00 and mode-11 transmission types. In both types, data are read at the receiver side at the rising edges of the clock pulses, and bits are placed onto the bus at the falling edges of the clock pulses. This is illustrated in Fig. 4.32.

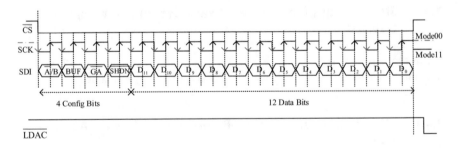

Fig. 4.32 SPI transmission waveforms for MCP4921

When $\overline{CS}$ is low, the four control bits and 12 data bits are samples at DAC at the rising edges of the clock pulses. When $\overline{CS}$ is raised high the data is latched onto the DAC's input registers. When $\overline{LDAC}$ pin achieves low state, the values held in the DAC's input registers are sent to the DCA's output registers.

In Fig. 4.32, it is seen that the first 4 bits are the configuration bits, and the next 12 bits are the data bits. The SPI frame, sent by the Master, consists of 16 bits. The first 4 bits are the configuration bits of the chip, and these bits are

$\overline{A}$ / B : channel select (used by MCP4922)
$\underline{BUF}$: VREF input buffer control bit
$\underline{GA}$: gain control
$\overline{SHDN}$: shutdown bit, turns off the output

and the other 12 bits represent the DAC value. The chip does not have MISO line as we mentioned before. An FPGA device can be connected to MCP4921. The VHDL implementation of the SPI protocol depicted in Fig. 4.32 for this device can be achieved easily as in Example 4.5.

4.4.2 Sine Signal Generation and SPI Protocol Development in VHDL for Digital Analog Converter (DAC), AD7303

In this section, we will consider the generation of sine signal in VHDL and implement the SPI protocol for digital-to-analog converter (DAC), AD7303. The generated sine signal is sent to AD7303. The output of the AD7303 can be observed on an oscilloscope screen.

We will explain the subject through examples. First, let us give some information about AD7303.

4.4.2.1 8-Bit Digital-to-Analog Converter, AD7303

The AD7303 is a dual port, 8-bit DAC that operates from a single +2.7 to +5.5 V supply. It has **mosi** pin and does not have **miso** pin as shown in Fig. 4.33. Maximum SPI communication speed is 30 MHz for AD7303. SPI mode utilized for the DAC chip is mode-00. Each transmission with AD7303 includes 8-bit control and 8-bit DAC data with a total of 16 bits.

The black-box representation of AD7303 is depicted in Fig. 4.33.

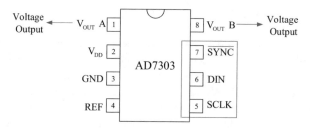

Fig. 4.33 AD7303 black-box representation

In Fig. 4.33, V_{OUT} A and V_{OUT} B are the output voltages. V_{DD} is the input power supply. REF is the reference input which ranges from 1 V to $V_{DD}/2$. SCLK is the serial clock. GND is the ground. DIN is the serial data input. AD7303 has 16-bit registers where 8 bits are used for data and 8 bits reserved for control operation. Data is read into the register at the rising edge of the clock input.

$\overline{\text{SYNC}}$ is used to control the input. When $\overline{\text{SYNC}}$ is low, data is loaded into the shift register at the rising edge of the clock pulses. The value of the output voltage at the terminals A or B equals to

$$V_0 = 2 \times V_{\text{REF}} \times \left(\frac{N}{256} \right)$$

where V_{REF} is the voltage at the REF pin or $V_{\text{DD}}/2$ when internal reference is chosen and N is the decimal equivalent of the 8-bit data, ranging from 0 to 255, loaded into DAC.

Data is sent into AD7303 with frames consisting of 16 bits of which the most significant 8 of them are used for control operation, and the least significant 8 bits are used for DAC. The control bits are used to select the outputs DAC A or DAC B, and they are used for various data loading functions, and for selecting between internal and external REF sources. For detailed information about the use of control bits, the reader can refer to the datasheet of AD7303.

Example 4.7 When the 8-bit data 11111111 is loaded into the registers of AD7303, the output voltage of the DAC can be calculated as

$$2 \times \frac{255}{256} \times V_{\text{REF}} \approx 2 \times V_{\text{REF}}$$

which is equal to V_{DD} when $V_{\text{REF}} = V_{\text{DD}}/2$. Note that V_{DD} ranges from 2.7 to 5.5 V.

4.4.2.2 Sine Signal Generation in VHDL
Next example illustrates how to generate a sine signal in VHDL.

Example 4.8 Write a VHDL code to generate sine signal.

Solution 4.8 We will consider the sine signal with frequency of 1 Hz, and generate the samples of 1 Hz sine signal for its one period, i.e., one cycle. Once we have the samples of 1 Hz sine signal for its one period, generation of any other sine signal with frequency f_0 Hz from an FPGA's output port can be achieved by transmitting f_0 cycles of sine signal through the port in 1 s. First, we will generate the samples of sine signal for its one cycle in MATLAB, and then use these samples in VHDL program.

The amplitude values of sine signal, $\sin(2\pi f_0 t)$ ranges from -1 to 1. We will generate 100 samples from one period of sine signal and these values will be represented by the integers in the range 0 to 255. An 8-bit DAC can handle integers ranging from 0 to 255.

In PR 4.26, 100 samples for sine signal for its one period are generated using MATLAB, and the generated values are mapped to integers ranging from 0 to 255, and the integer values are written into a file. The frequency of the sine signal is set to 1 Hz.

```
clc; clear all;
t=1:100;
f=1;
y=sin(2*pi*f*t/100)*127+128;
plot(t,y)
y=round(y);
fprintf('SINROM<=(')
fprintf('%.0f,' , y(1:end-1));
fprintf('%.0f' , y(end));
fprintf(')')
```

PR 4.26   Program 4.26

The integer sine samples obtained from PR 4.26 are depicted in PR 4.27.

```
y=[136, 144, 152, 160, 167, 175, 182, 189, 196, 203, 209, 215, 221, 226, 231, 235,
239, 243, 246, 249, 251, 253, 254, 255, 255, 255, 254, 253, 251, 249, 246, 243, 239,
235, 231, 226, 221, 215, 209, 203, 196, 189, 182, 175, 167, 160, 152, 144, 136, 128,
120, 112, 104, 96, 89, 81, 74, 67, 60, 53, 47, 41, 35, 30, 25, 21, 17, 13, 10, 7, 5, 3, 2,
1, 1, 1, 2, 3, 5, 7, 10, 13, 17, 21, 25, 30, 35, 41, 47, 53, 60, 67, 74, 81, 89, 96, 104,
112, 120, 128]
```

PR 4.27   Program 4.27

When the integer vector in PR 4.27 is plotted using MATLAB, we obtain the graph shown in Fig. 4.34.

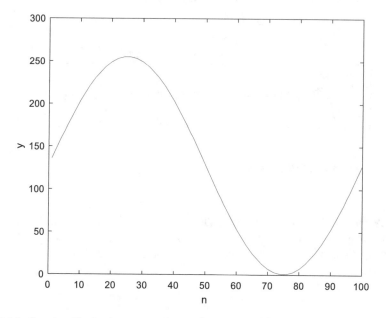

Fig. 4.34   Sine signal having integer amplitude values

The MATLAB generated sine samples can be used in a VHDL program as illustrated in PR 4.28.

```
type sinlut is array(0 to 99) of integer range 0 to 255;
signal sinrom: sinlut;
begin
sinrom<=(136, 144, 152, 160, 167, 175, 182, 189, 196, 203, 209, 215,
        221, 226, 231, 235, 239, 243, 246, 249, 251, 253, 254, 255, 255,
        255, 254, 253, 251, 249, 246, 243, 239, 235, 231, 226, 221, 215,
        209, 203, 196, 189, 182, 175, 167, 160, 152, 144, 136, 128, 120,
        112, 104, 96, 89, 81, 74, 67, 60, 53, 47, 41, 35, 30, 25, 21, 17, 13,
        10, 7, 5, 3, 2, 1, 1, 1, 2, 3, 5, 7, 10, 13, 17, 21, 25, 30, 35, 41, 47,
        53, 60, 67, 74, 81, 89, 96, 104, 112, 120, 128);
```

PR 4.28   Program 4.28

Example 4.9 AD7303 is a serial input, dual channel output 8-Bit DAC. Generate sine signal in FPGA using VHDL and transmit the sine samples from FPGA to AD7303 through SPI port.

Assume that the output of the AD7303 is connected to an oscilloscope where sine signal is observed.

Solution 4.9 The connections between FPGA, DAC, and scope are shown in Fig. 4.35 where it is seen that there is no MISO line.

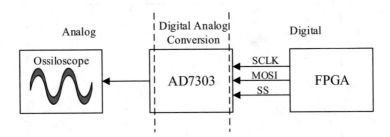

Fig. 4.35 FPGA to AD7303 SPI connections

FPGA controls DAC through 3-wire SPI line. We need to send control commands and sine signal samples. AD7303 supports Mod-00 SPI communication. In the datasheet of AD7303, SPI communication waveforms of AD7303 are drawn as in Fig. 4.36.

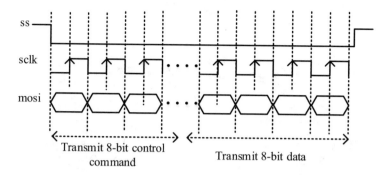

Fig. 4.36   Mod-00 SPI communication for AD7303

The data to be transmitted are the 8-bit sine samples. The sine samples can be generated using MATLAB as in the previous example and can be directly used in VHDL code.

There are 100 sine samples generated, and each 8-bit sample is transmitted with an 8-bit control command as depicted in Fig. 4.37 where it is seen that 16 bits are transmitted in each transmission session, consisting of 8-bit control and 8-bit data integers, and between two transmission sessions "**ss**" signal is set to "1". There are 100 transmission sessions.

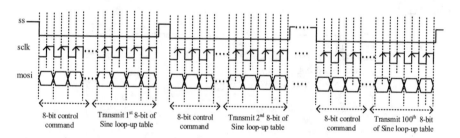

Fig. 4.37   Transmission of 100 sine samples using SPI protocol for AD7303

AD7303 SPI serial clock frequency can have values up to 30 MHz. In our application, we choose the serial clock frequency of the SPI as 1 MHz. FPGA's clock frequency is accepted as 100 MHz. The serial clock of SPI is obtained from the clock of FPGA using a frequency divider process. The state diagram of data transmission via SPI communication can be drawn as in Fig. 4.38.

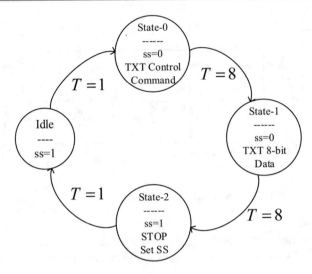

Fig. 4.38 State diagram of SPI protocol for AD7303

The entity unit and declarative part of the architecture are given in PR 4.29. In entity part, SPI port signals are defined. In the declarative part of the architecture unit, the signal object "count", to be used for frequency division operation, is defined. The signal object "sinrom" is used to store the sine sample values generated by MATLAB.

```
library ieee;
use ieee.std_logic_1164.all;
use ieee.std_logic_arith.all;
use ieee.std_logic_unsigned.all;

entity AD7303_driver is
  port(clk_100MHz, reset: in std_logic;
       ss, mosi, sclk: out std_logic);
end AD7303_driver;

architecture logic_flow of AD7303_driver is
  type state is (st_idle, st_DacSent, st_Stop);
  signal present_state, next_state: state;
  signal spi_sclk: std_logic;
  signal count: positive range 1 to 500:=1;
  signal data: std_logic_vector(15 downto 0);
  signal control: std_logic_vector(7 downto 0):="00000000";
  type sinlut is array(0 to 99) of integer range 0 to 255;
  signal sinrom: sinlut;
  constant data_length: natural:=16;
  constant max_length: natural:=15;
  signal data_index: integer range 0 to 15;
  signal timer: natural range 0 to data_length;
  signal sample_index : integer range 0 to 100:=0;
```

PR 4.29 Program 4.29

In PR 4.30, sine sample values are assigned to "sinrom" signal object, and generation of 1 MHz serial clock of SPI bus is achieved using a frequency divider process.

```
begin
 sinrom<=(136, 144, 152, 160, 167, 175, 182, 189, 196, 203, 209, 215,
         221, 226, 231, 235, 239, 243, 246, 249, 251, 253, 254, 255, 255,
         255, 254, 253, 251, 249, 246, 243, 239, 235, 231, 226, 221, 215,
         209, 203, 196, 189, 182, 175, 167, 160, 152, 144, 136, 128, 120,
         112, 104, 96, 89, 81, 74, 67, 60, 53, 47, 41, 35, 30, 25, 21, 17, 13,
         10, 7, 5, 3, 2, 1, 1, 1, 2, 3, 5, 7, 10, 13, 17, 21, 25, 30, 35, 41, 47,
         53, 60, 67, 74, 81, 89, 96, 104, 112, 120, 128);

 p1: process(clk_100MHz, reset)
 begin
  if(reset='1') then
    count<= 1;
  elsif(rising_edge(clk_100MHz)) then
    count<=count+1;
    if(count=50) then
     spi_sclk<=not spi_sclk;
     count<=1;
    end if;
   end if;
  end process;
  sclk<=spi_sclk;
```

PR 4.30 Program 4.30

SPI protocol can be implemented using the processes in PR 4.31.

```
p2: process(spi_sclk, reset)
begin
 if(reset='1') then
  present_state<=st_idle;
  data_index<=0;
  sample_index<=0;
 elsif(spi_sclk'event and spi_sclk ='0') then
  if(data_index=timer-1) then
   present_state<=next_state;
   data_index<=0;
   if(timer>1)then
    if(sample_index=99) then
     sample_index<= 0;
    else
     sample_index<=sample_index +1;
    end if;
 data<= control&conv_std_logic_vector(sinrom(sample_index),8);
   end if;
  else
   data_index<=data_index +1;
  end if;
 end if;
end process;
```

```
p3: process(present_state)
begin
 case present_state is
  when st_idle =>
   ss<='1';
   mosi<='X';
   timer<=1;
   next_state<=st_DacSent;
  when st_DacSent =>
   ss<='0';
   timer<=16;
   mosi<=data(15-data_index);
   next_state<=st_Stop;
  when st_Stop =>
   ss<='1';
   mosi<='X';
   timer<=1;
   next_state<=st_idle;
  end case;
 end process;
end logic_flow;
```

PR 4.31 Program 4.31

For the transmission of a single sine sample, we need 8 clock cycles for the transmission of control command, another 8 clock cycles for the transmission of sine sample, and 2 clock cycles for the stop and idle states. In total, we need $8 + 8 + 2 = 18$ clock cycles for the transmission of a single sine sample. The frequency of sine signal generated by VHDL can be calculated as

$$f = \frac{1\,\text{MHz}}{100 \times 18} \rightarrow 555.555\,\text{Hz}$$

The sine signal generated using PR 4.31 is viewed on the screen of a digital oscilloscope as in Fig. 4.39.

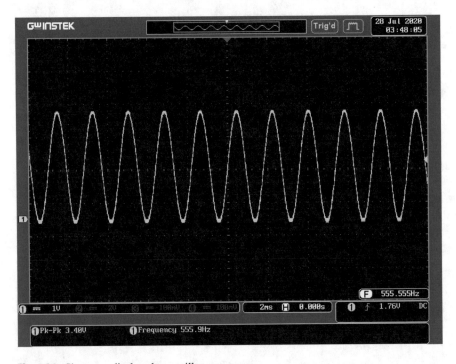

Fig. 4.39 Sine wave displayed on oscilloscope screen

4.4.2.3 Second Solution

Instead of using two separate process, we can achieve the implementation using a single process as in PR 4.32.

```vhdl
library ieee;
use ieee.std_logic_1164.all;
use ieee.std_logic_arith.all;
use ieee.std_logic_unsigned.all;
entity AD7303_driver is
  port (clk_100MHz, reset: In std_logic;
        ss, mosi, sclk: out std_logic);
end AD7303_driver;


architecture logic_flow of AD7303_driver is

  type state is (idle_st, dac_st);
  signal present_state: state:=idle_st;
  signal clk_1MHz: std_logic:='0';
  signal count: positive range 1 to 50:=1;
  signal data: std_logic_vector(15 downto 0);
  type sinlut is array(0 to 99) of integer range 0
               to 255;
  signal sinrom: sinlut;
begin
  sinrom<=(136, 144, 152, 160, 167, 175, 182,
           189, 196, 203, 209, 215, 221, 226,
           231, 235, 239, 243, 246, 249, 251,
           253, 254, 255, 255, 255, 254, 253,
           251, 249, 246, 243, 239, 235, 231,
           226, 221, 215, 209, 203, 196, 189,
           182, 175, 167, 160, 152, 144, 136,
           128, 120, 112, 104, 96, 89, 81, 74,
           67, 60, 53, 47, 41, 35, 30, 25, 21, 17,
           13, 10, 7, 5, 3, 2, 1, 1, 1, 2, 3, 5, 7,
           10, 13, 17, 21, 25, 30, 35, 41, 47, 53,
           60, 67, 74, 81, 89, 96, 104, 112, 120,
           128);

  p_cdiv: process(clk_100MHz, reset)
  begin
    if(reset='1') then
      count<=1;
    elsif(rising_edge(clk_100MHz)) then
      count<=count+1;
      if(count=50) then
        clk_1MHz<=not clk_1MHz;
        count<=1;
      end if;
    end if;
  end process;

  sclk<= clk_1MHz;

  p1_p2: process(clk_1MHz, reset)
    variable bit_index: integer range 0 to 16:=0;
    variable sample_index: integer range 0 to
                                    100:=0;
  begin
    if(reset='1') then
      ss<='1';
      present_state<=idle_st;
    elsif(falling_edge(clk_1MHz)) then
      case present_state is
        when idle_st =>
          ss<='1';
          present_state<=dac_st;

        when dac_st =>
          if(bit_index=16) then
            ss<= '1';
            bit_index:=0;
            sample_index:=sample_index + 1;
            if(sample_index = 100) then
              sample_index:=0;
            end if;
            data<=x"00"&conv_std_logic_vector
                          (sinrom(sample_index),8);
            present_state<=dac_st;
          else
            ss<='0';
            mosi<=data(15-bit_index);
            bit_index:=bit_index+1;
            present_state<=dac_st;
          end if;

        when others =>

      end case;
    end if;
  end process;

end logic_flow;
```

PR 4.32 Program 4.32

4.4.3   SPI Protocol Development in VHDL for Digital Output MEMS Accelerometer, ADXL362

The ADXL362, which works at exceptionally low power consumption levels, is a complete 3-axis acceleration measurement device. It measures both dynamic acceleration and static acceleration. Motion or shock yields dynamic acceleration, and tilt cause static acceleration. Acceleration is informed digitally, and the device communicates via the SPI protocol.

The black-box representation of the ADXL362 is depicted in Fig. 4.40 where SPI pins are shown in red color.

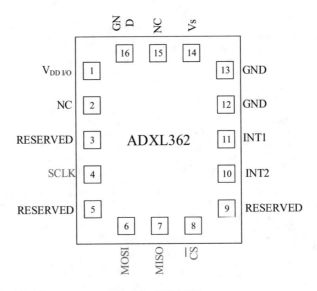

Fig. 4.40   The black-box representation of the ADXL362

4.4.3.1 SPI Protocol of ADXL362

There are five different types of SPI communications that ADXL362 supports. These communication types are "Register Read", "Register Write", "Burst Read", "Burst Write", and "FIFO Read". In all types of SPI communications, first instruction is sent, then address is sent, and lastly data is sent or received.

In this section, we will only consider "Register Read" SPI communication type of ADXL362. For the other SPI communication types, the reader can refer to the datasheet of ADXL362. SPI waveforms corresponding to "Register Read" type of communication is shown in Fig. 4.41 where it is seen that first "read instruction", followed by the transmission of register address to be read, is sent, and then 8-bit data is received. Each of "Read instruction", "Register address", and "Read data" contains 8-bit.

The "Read instruction/command" for SPI protocol of ADXL362 is the 8-bit string 00001011.

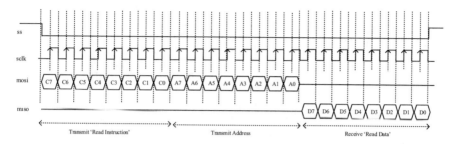

Fig. 4.41 SPI waveforms for ADXL362

Now, we consider the development of SPI protocol in VHDL so that data communication between ADXL362 and FPGA can be achieved. We will explain the subject through an example.

Example 4.10 ADXL362 is a 3-Axis digital output MEMS accelerometer that communicates through SPI. Read 8-bit device identification (ID) of the chip and show this device ID on LEDs of your FPGA board. Figure 4.42 shows the master, FPGA, and slave, ADXL362, connections.

Fig. 4.42 SPI lines between ADXL362 and FPGA

Solution 4.10 When the register map of the ADXL362 is inspected device, it is seen that ID is put in 0x00 address by the manufacturer. Its default value is hexadecimal 0xAD. SPI read from ADXL362 can be accomplished in three steps.

1. Set SS to "0" and send 8-bit Read command through MOSI line.
2. Send 8-bit Address information through MOSI line.
3. Read data in this address through MISO line and set SS to "1".

Timing waveforms for serial communication are shown in Fig. 4.43. After equating SS to 0, read command, i.e., 0x0B, is transmitted starting from its the most significant bit. Read command bits are denoted by C7, C6, ..., C0, and its binary equivalent is 00001011. In the second step, address bits, 00000000, denoted by A7, A6, ..., A0 are transmitted. In the last step, slave device, ADXL362, sends the data stored in the 0x00 register via MISO line. The data transmitted is the device ID of the ADXL362 which is 0xAD.

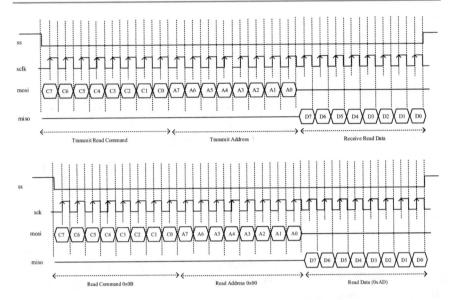

Fig. 4.43 SPI protocol data receive and transmit operations for ADXL362

In PR 4.33, ports are defined in the entity part, and in the declarative part of the architecture signal object definitions are made, and **state** data type is introduced. The value of the signal object "rdid" contains the "read" command 0 × 0B and device ID address 0 × 00. "spi_sclk" and "count" signal objects are to be used for clock division operation. The clock frequency of SPI communication is 2 MHz.

```
library ieee;                                          p1:process(clk_100MHz,reset)
use ieee.std_logic_1164.all;                           begin
use ieee.numeric_std.all;                                if (falling_edge(clk_100MHz)) then
entity adxl362_read is                                      count<=count+1;
  port (clk_100MHz,reset in  std_logic;                    if (count=24) then
       miso:  in  std_logic;                                  spi_sclk<=not spi_sclk;
       cs,mosi,sclk; out  std_logic;                          count<=0;
       led: out  std_logic_vector (7 downto 0));          end if;
end adxl362_read;                                        end if;
                                                       end process;
architecture logic_flow of adxl362_read is
  signal rdid: std_logic_vector(15 downto 0):=x"0B00";
  type state is (st_idle,st_rxmt,st_read,st_stop);
  signal present_state, next_state: state;
  constant data_length: natural:=24;
  signal timer: natural range 0 to data_length;
  signal data_index: natural range 0 to data_length;
  signal spi_sclk,transition_done:std_logic;
  signal count : integer range 0 to 50:=0;
  signal read_data : std_logic_vector(7 downto 0);
  begin
```

PR 4.33 Program 4.33

There are three processes written in PR 4.34. The process "p2" is used to update the present state. The process "p3" is used to receive the data sent by slave, and the last process "p4" is used to transmit data, which contains "read" command and address information, from master to slave.

```vhdl
led<=read_data;
sclk<= spi_sclk when transition_done = '0' else '0';

p2: process(spi_sclk, reset)
begin
  if(reset='1') then
    present_state<=st_idle;
    data_index<=0;
  elsif(spi_sclk'event and spi_sclk='0') then
    if(data_index=timer-1) then
      present_state<=next_state;
      data_index<=0;
    else
      data_index <=data_index +1;
    end if;
  end if;
end process;
p3: process(spi_sclk)
begin
  if(spi_sclk'event and spi_sclk='1') then
    if(present_state=st_read) then
      read_data(7-data_index)<=miso;
    end if;
  end if;
end process;
```

```vhdl
p4: process(present_state, data_index)
begin
  case present_state is
    when st_idle =>
      transition_done<='0';
      cs<='1';
      mosi<='X';
      timer<=1;
      next_state<=st_rxmt;
    when st_rxmt =>
      cs<='0';
      timer<=16;
      mosi<=rdid(15-data_index);
      next_state<=st_read;
    when st_read =>
      cs<='0';
      timer<=8;
      next_state<=st_stop;
    when st_stop =>
      cs<='1';
      timer<=1;
      transition_done<='1';
      next_state<=st_stop;
  end case;
end process;
```

PR 4.34 Program 4.34

Combining all the program parts, we get the overall implementation.

Example 4.11 ADXL362 is a 3-Axis digital output MEMS accelerometer that communicates with SPI protocol. First, write the data 0xBC to the register address 0x20 of ADXL362, and then read the content of this register address information and show this content on LEDs of your FPGA kit.

Solution 4.11 Writing data to ADXL362's register can be achieved via the steps

1. Reset SS to "0" and send 8-bit write command through MOSI line.
2. Send 8-bit address information (0x20) through MOSI line.
3. Send 8-bit data (0xBC) through MOSI line and Set SS to "1".

After write operation, read operation is performed. The read operation can be achieved through the steps.

1. Reset SS to "0" and send 8-bit read command through MOSI line.
2. Send 8-bit address information (0x20) through MOSI line.
3. Read data that comes from MISO line and set SS to "1".

The timing waveforms for write and read operations are illustrated in Fig. 4.44.

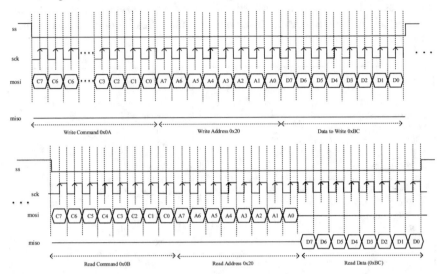

Fig. 4.44   SPI protocol waveforms for Example 4.11

Considering the waveforms in Fig. 4.44, we write the entity part and define signal objects in the declarative part of the architecture as in PR 4.35.

```
library ieee;
use ieee.std_logic_1164.all;
use ieee.numeric_std.all;

entity adxl362_write_read is
  port (clk, reset: in std_logic;
        cs, mosi, sclk: out std_logic;
        miso: in std_logic;
        led: out std_logic_vector (7 downto 0));
end adxl362_write_read;

architecture logic_flow of adxl362_write_read is

  signal wrid: std_logic_vector(23 downto 0):=x"0A20BC";
  signal rdid: std_logic_vector(15 downto 0):=x"0B20";
  type state is (st_idle, st_txmt, st_stop1, st_rxmt, st_read, st_stop2);
  signal present_state, next_state: state;
  constant data_length: natural:=24;
  signal timer: natural range 0 to data_length;
  signal data_index: natural range 0 to data_length;
  signal spi_sclk, transition_done: std_logic;
  signal count: integer range 0 to 50:=0;
  signal read_data: std_logic_vector(7 downto 0);
begin
```

PR 4.35   Program 4.35

The process written for the generation of 2 MHz clock frequency is given in PR 4.36.

```
p1: process(clk, reset)
begin
  if(falling_edge(clk)) then
    count<=count+1;
    if(count=24) then
      spi_sclk<=not spi_sclk;
      count<=0;
    end if;
  end if;
end process;
```

PR 4.36 Program 4.36

There are three processes written in PR 4.37 to send and receive data with SPI protocol. The process "p2" is used to update the present state. The process "p3" is used to receive the data sent by slave. The last process "p4" is used to transmit "write", "read" commands, and "address" and "data" bits.

```
led<=read_data;
sclk<=spi_sclk when transition_done='0' else '0';

p2: process(spi_sclk, reset)
begin
  if(reset='1') then
    present_state<=st_idle;
    data_index<=0;
  elsif(spi_sclk'event and spi_sclk='0') then
    if(data_index=timer-1) then
      present_state<=next_state;
      data_index<=0;
    else
      data_index<=data_index +1;
    end if;
  end if;
end process;
p3: process(spi_sclk)
begin
  if(spi_sclk'event and spi_sclk='1') then
    if(present_state=st_read) then
      read_data(7-data_index)<=miso;
    end if;
  end if;
end process;

p4: process(present_state, data_index)
begin
  case present_state is
    when st_idle =>
      transition_done<='0';
                                    cs<='1';
                                    mosi<='X';
                                    timer<=1;
                                    next_state<=st_txmt;
                                  when st_txmt =>
                                    cs<='0';
                                    timer<=24;
                                    mosi<=wrid(23-data_index);
                                    next_state<=st_stop1;
                                  when st_stop1 =>
                                    cs<='1';
                                    timer<=1;
                                    next_state<=st_rxmt;
                                  when st_rxmt =>
                                    cs<='0';
                                    timer<=16;
                                    mosi<=rdid(15-data_index);
                                    next_state<=st_read;
                                  when st_read =>
                                    cs<='0';
                                    timer<=8;
                                    next_state<=st_stop2;
                                  when st_stop2 =>
                                    cs<='1';
                                    timer<=1;
                                    transition_done<='1';
                                    next_state<=st_stop2;
                                  end case;
                                end process;
                                end logic_flow;
```

PR 4.37 Program 4.37

Exercise AD9528 is a clock generator that is used for reference clock sources in high-performance wireless transceivers, LTE and multicarrier GSM base stations, wireless and broadband infrastructure, and medical instrumentation. It supports both 3-wire and 4-wire SPI communication protocols. Refer to the datasheet of AD9528 and write a VHDL program for 3-wire SPI mode to read chip's Vendor ID and show the result on LEDs of your FPGA kit.

Problems

1. Draw the SPI waveforms for the transmission of bit sequence 0111011 in mode-00 transmission scheme and implement it in VHDL.
2. The state diagram of an SPI protocol involving transmit and receive operations is depicted in Fig. P4.1. Implement SPI protocol diagram in VHDL. Use a serial bus clock of 1 MHz for SPI protocol. Use dummy values for the transmit and received data bits.

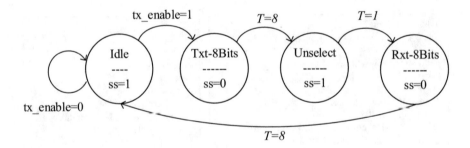

Fig. P4.1 State diagram for P2

3. In synchronous communication, both receive and transmit operations are performed at the rising edge of clock pulses. Is this statement correct or not? If it is not correct, then give an example violating the statement in the sentence.
4. What is the difference between Mod-00 and Mod-01 SPI transmission operation modes.
5. What is the maximum clock frequency for SPI communication?
6. What does SPI mean?
7. Refer to the datasheet of AD9528 and decide which SPI operation mode is supported by SPI protocol of AD9528.
8. List several electronic components having SPI communication ports.
9. How many process units are available in a VHDL implementation of SPI protocol if the protocol handles only transmit operation.
10. How many process units are available in a VHDL implementation of SPI protocol if the protocol handles both transmit and receive operations.

Inter Integrated Circuit (I2C) Serial Communication in VHDL

5

Inter integrated circuit (I2C) is a synchronous serial communication protocol. It is developed for serial communication between electronic devices. I2C is an 8-bit oriented communication protocol. I2C communication employs two wires for communication. I2C synchronous communication protocol has been developed by the Philips company in 1982. It is widely used for connecting low-speed peripheral devices to more complex electronic devices such as processors and microcontrollers for short-distance or intra-board communications. Several versions of I2C protocol have been released in time. In 2007, the version 3.0 was introduced. Version 4, which adds 5 Mbit/s ultra fast-mode, is introduced in 2012. In 2014, the last revision, version 6, was released. In this chapter, we first explain the I2C communication, then provide information about VHDL implementation of I2C communication protocol. We further explain the subject using clear examples.

5.1 Master-Slave Connections and I2C Port Circuit

A typical I2C bus is depicted in Fig. 5.1 where a single master which is usually a microcontroller drives a number of slaves which are electronic devices named as peripherals. The peripherals can be sensors, EEPROMs, LCDs, and other microcontrollers as well. In Fig. 5.1, "scl" corresponds to serial clock, and "sda" corresponds to serial data.

In I2C communication every peripheral has an assigned device address. Whenever the master wants to transmit some data to a peripheral, it first issues the address of the device onto the bus and then transmits the data targeted to the device.

© The Author(s), under exclusive license to Springer Nature Switzerland AG 2021
O. Gazi, A. Ç. Arlı, *State Machines using VHDL*,
https://doi.org/10.1007/978-3-030-61698-4_5

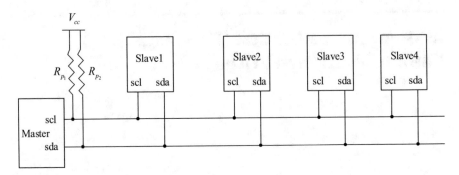

Fig. 5.1 Master connected to several slaves

As it is seen from Fig. 5.1 that pull-up resistors are connected to the SCL and SDA lines. The values of the pull-up resistors are calculated considering the capacitance of SCL and SDA wires. Typical values of pull-up resistors range from 1 to 47 kΩ. Pull-up resistors are used due to the open-drain structure of the SCL/SDA port circuits. The port circuit structure of SCL and SDA is depicted in Fig. 5.2.

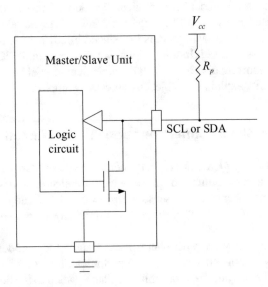

Fig. 5.2 SLC or SDA port circuit

Master and slave devices are connected as in Fig. 5.3 where instead of two lines a single line is drawn for the simplicity of illustration.

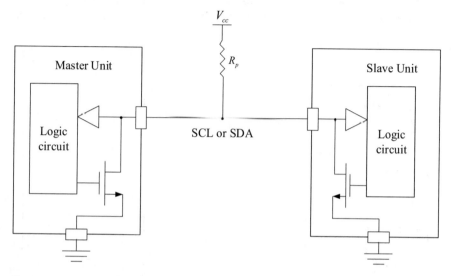

Fig. 5.3 Master-slave connection

When master (or slave) wants to transmit logic-0, it activates the transistor shorting the port output to the ground, and the receiver releases the bus making the open-drain collector output high impedance, i.e., Z logic as shown in Fig. 5.4.

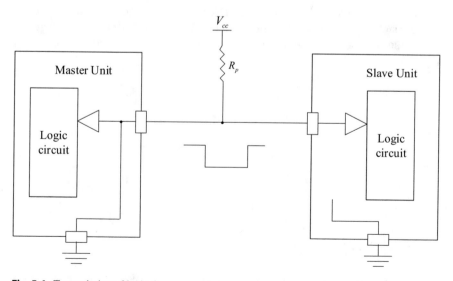

Fig. 5.4 Transmission of logic-0

When master (or slave) wants to transmit the logic-1, it releases the bus by turning off the transistor. This leaves the bus floating and the pull-up resistor raises the voltage of the bus to the V_{cc} level, and this will be interpreted as the high logic. This is illustrated in Fig. 5.5 where master transmits logic-1 to slave which acts as receiver.

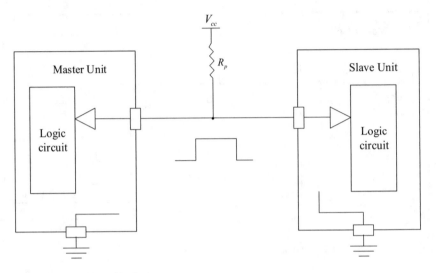

Fig. 5.5   Transmission of logic-1

5.2   START, STOP, and IDLE Control Signals of I2C Protocol

The I2C bus is "Idle" if both SCL and SDA lines are at high logic. To initiate the data transmission, a start signal depicted in Fig. 5.6 is sent. The start signal is generated when SDA makes a high to low transition while SCL is high as it is seen in Fig. 5.6.

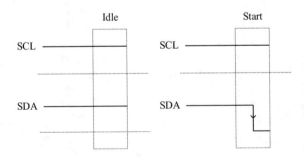

Fig. 5.6   "Idle" and "start" signals

After the transmission of start signal, data transmission takes place followed by a "Stop" signal which is depicted in Fig. 5.7.

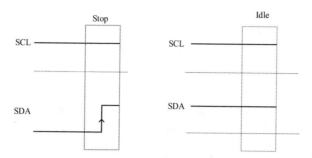

Fig. 5.7 "Stop" and "idle" signals

The data transmission takes place after sending the start signal, and when the transmission of the last bit is complete, the stop signal is issued as indicated in Fig. 5.8.

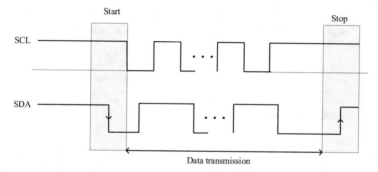

Fig. 5.8 Data transmission between "start" and "stop" signals

5.3   Generation of Shifted Clock and Determination of the Transmission Instants

The information bits are placed onto the SDA line when SCL is low. It is important to note that new bits are never placed onto the bus when SCL is high. This concept is illustrated in Fig. 5.9.

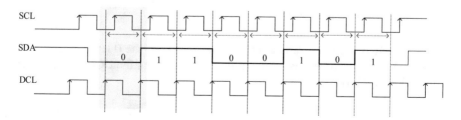

Fig. 5.9 I2C transmission waveforms

In Fig. 5.9, we also drew a reference clock denoted by DCL, called data-clock, which is a shifted version of the SCL to indicate the time instants at which transmission of the information bits over SDA line occur. As it is clear from Fig. 5.9 that new information bits are placed onto the SDA line at the rising edges of the data-clock, DCL.

Master device can transmit data to the slave, and slave can also transmit data to the master. The transmission of the data from master to slave is named as the write operation, and the transmission of data from slave to master is named as the read operation. To comprehend the subject better, let us solve an example to illustrate the use of timing waveforms for I2C communication.

Example 5.1 Draw the timing waveforms of the I2C protocol for the transmission of the bit stream 10001101.

Solution 5.1 Let us first draw a clock waveform including ten pulses from which eight of them will be utilized for the data transmission, and the other two will be used for the START and STOP signaling. The clock waveform is depicted in Fig. 5.10.

SCL

Fig. 5.10 Serial clock waveform

In the next step, we draw data-clock, DCL, waveform such that if the pulse with of the SCL waveform is T, the reference clock waveform is shifted to the left by $T/4$. The data-clock waveform together with the original clock waveform is shown in Fig. 5.11.

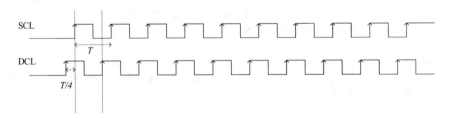

Fig. 5.11 Generation of data-clock

In the next step, we draw vertical lines to the centers of the 0 levels of the SCL waveform, and these lines pass through the rising edges of the DCL waveform as depicted in Fig. 5.12.

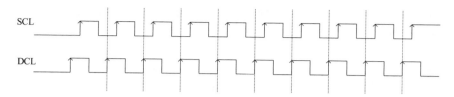

Fig. 5.12 The relationship between serial clock and data-clock

DCL waveform can be obtained from SCL clock waveform and START signal can be chosen as the first pulse of the DCL clock as shown in Fig. 5.13.

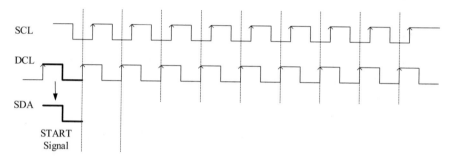

Fig. 5.13 Generation of "start" signal

Following the START signal, the transmissions of data bits take place. The most significant bit of the data byte which is "1" for our example is transmitted as in Fig. 5.14 where it is seen that the bit "1" is transmitted at the rising edge of the data-clock, and it has one clock duration, and the transmission duration corresponds to the mid-points of the low-logic of the SCL waveform.

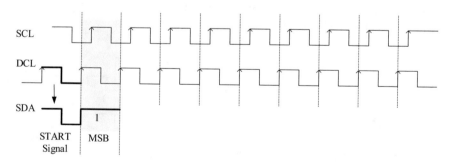

Fig. 5.14 Transmission of MSB

The next bit to be transmitted is "0", and the transmission waveform for this bit is depicted in Fig. 5.15.

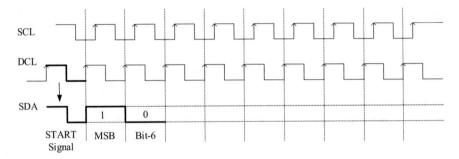

Fig. 5.15 Transmission of bit-6

Considering the transmission of the other information bits, we obtain the transmission waveform of Fig. 5.16.

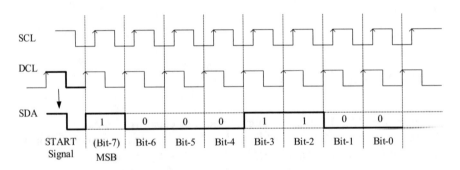

Fig. 5.16 I2C data transmission

After the transmission of last data bit, stop signal, which is formed by taking the complement of the DCL, is sent as shown in Fig. 5.17.

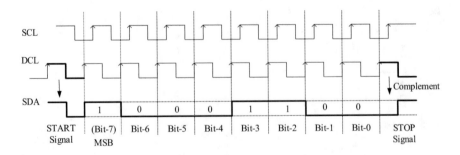

Fig. 5.17 Obtaining "stop" signal from data-clock

5.4   I2C Read and Write Operations

In this section, we provide more information about the I2C read and write operations.

5.4.1   I2C Write Operation

As it is mentioned before, data transmission from master to slave is called "write" operation. The write operation can be performed using the steps:

1. Master transmits the START signal.
2. Master places the device address concatenated with WRITE flag and data onto the SDA line.
3. Master halts the transmission sending the STOP signal.

5.4.2   I2C Read Operation

If master requests some data from slave, the set of operations until slave data is received by master is called read operation. Read operation is a more involved process. The read operation for I2C protocol can be summarized as:

1. Master transmits the START signal.
2. Master places the device address and data onto the SDA line.
3. If the peripheral needs some information for the data to be transmitted to the master, the master first sends some commands each followed by an acknowledgment (ACK) signal sent by the slave. Then, the master sends the START signal and device address again, and the slave transmits the data requested by the master. When the reading is complete, the master sends the STOP signal again.

In the second type of read operation, the slave does not need any command. The master sends the device's address with READ flag after START signaling, then the slave transmits the data needed by master followed by the STOP signal. Reading temperature from a sensor can be given as an example to this type of I2C communication.

Example 5.2 Reading data from an EEPROM can be given as an example. The master first transmits START signal, then transmits the device address with WRITE flag, then transmits the register address to be read, and gets the ACK signal from the slave. Next, the master sends the START signal and device address with READ flag, and it reads the data transmitted by the slave until it terminates the reading operation sending the STOP signal.

Considering the above discussion, we explain standard I2C data transfer formats as follows.

5.5   Data Transfer Formats

In this section, we explain the data transfer formats available in the datasheet of I2C standard. The slave peripherals can have 7-bit or 10-bit addresses. The frame formats explained in this section are used when slaves use 7-bit address. For 10-bit addressing frame formats, we advise the reader to refer to "UM10204, I2C-bus specification and user manual".

5.5.1   Write Operation

As it is stated before, data transmission from master to the slave receiver is also called write operation. The data transfer format of the write operation is depicted in Fig. 5.18.

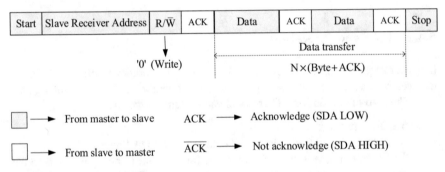

Fig. 5.18 Frame format for "write" operation

In Fig. 5.18, it is seen that each received byte is acknowledged by the slave. In Fig. 5.18, although we only showed ACK signals in the frame format, $\overline{ACK}$ signals can also be sent by the slave.

5.5.2   Master Reads the Slave Immediately

The master reads the data transferred by the slave immediately after issuing the slave address and read request flag. The data transfer format of this type of communication is depicted in Fig. 5.19.

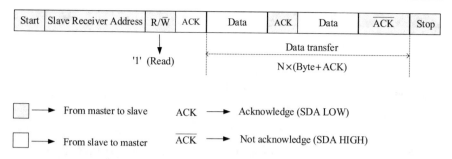

Fig. 5.19 Frame format for immediate "read" operation

It is seen from Fig. 5.19 that each byte is acknowledged by the master except the last one just before the STOP signal. The master sends negative-acknowledge, $\overline{ACK}$, just before the STOP signal.

5.5.3   Combined Format Involving Repeated START

In combined format, if a change of direction within a transfer occurs, then the START signal and the slave address are both sent again, but with the R/W bit reversed. If a master-receiver sends a repeated START signal, it sends negative-acknowledge, $\overline{ACK}$, just before the repeated START signal.

The most general form of the combined format is depicted in Fig. 5.20 where it is seen that the master first performs a read/write operation followed by ACK signal from slave, then data transfer takes place, and the steps are repeated following a second start signal till the issue of the STOP signal.

Fig. 5.20 Frame format for combined read-write operations

A transmission scheme involving a START signal immediately followed by a STOP signal does not comply with a legal frame format.

Example 5.3 A specific instance of the data transmission format in Fig. 5.20 can be given as an example in Fig. 5.21 where reading data from an EEPROM by a microcontroller via I2C protocol is demonstrated.

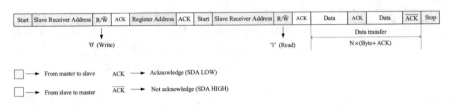

Fig. 5.21 Frame format for Example 5.3

In Fig. 5.21, it is seen that the master first performs a write operation, then it performs a read operation.

Example 5.4 In Fig. 5.22, the master first performs a read operation, then in sequel it performs a write operation, i.e., a read operation is followed by a write operation.

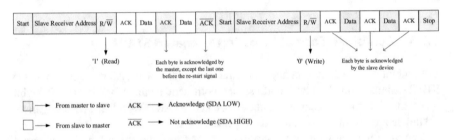

Fig. 5.22 Frame format for Example 5.4

Example 5.5 Draw the state diagram of I2C communication protocol where a master device sends only an 8-bit data to a slave device. Use all zero-bit sequences for 7-bit slave address.

Solution 5.5 This is a simple write operation. The frame structure for this operation is depicted in Fig. 5.23.

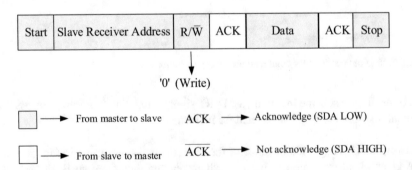

Fig. 5.23 Frame format for Example 5.5

The transmission frame format in Fig. 5.23 can be described using Moore state diagram as in Fig. 5.24.

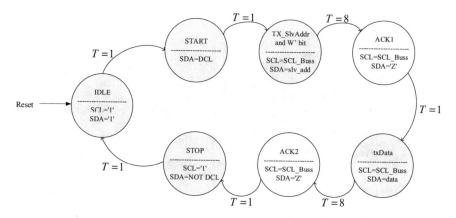

Fig. 5.24 State diagram for Fig. 5.23

5.6 VHDL Implementation of I2C Protocol

The VHDL implementation of I2C protocol can be achieved using the timed state machines. However, before proceeding to the implementation of I2C protocol via timed state machines, it is useful to study the implementation of some program units which are used in VHDL implementation of timed state machines. Considering the bus speed of the I2C protocol and clock frequency of FPGA device, it is clear that we need a clock divider to generate the bus frequency, i.e., SCL clock frequency from FPGA clock frequency. In addition, we need a data-clock which is obtained by shifting bus-clock SCL.

Once we have the data-clock, DCL, we can generate the START and STOP signals used at the beginning and at the end of SDA waveform. Assume that SCL frequency is 1 MHz, and FPGA's clock frequency is 100 MHz. To generate the 1 MHz SCL signal, we first generate 4 MHz clock waveform, and using 4 MHz clock we can generate 1 MHz I2C bus-clock SLC and data-clock DCL. We can design a frequency divider that generates a clock frequency of 4 MHz from a clock frequency of 100 MHz as follows.

Using the formula

$$\frac{f}{2K}$$

for the desired frequency, i.e.,

$$\frac{f}{2K} = 4\,000\,000$$

and substituting $f = 100\,000\,000$, we get $K = 12$. Thus, using an 0 to 10 counter, we can approximately achieve the desired clock frequency as in PR 5.1.

```vhdl
library ieee;
use ieee.std_logic_1164.all;

entity buss_clk_gen is
  port (clk_100MHz: in std_logic;
        scl: out std_logic);
end entity;

architecture logic_flow of buss_clk_gen is
  signal count: natural range 0 to 11:=0;
  signal clk_4MHz: std_logic:='0';


begin
```

```vhdl
process(clk_100MHz)
begin
  if(rising_edge(clk_100MHz)) then
    count<=count + 1;
    if(count=11) then
      clk_4MHz<=not clk_4MHz;
      count<=0;
    end if;
  end if;
end process;

end architecture;
```

PR 5.1 Program 5.1

We can generate 1 MHz bus-clock SCL and 1 MHz reference clock DCL using the process unit in PR 5.2.

```vhdl
clk1MHz: process (clk_4MHz)
  variable count_1: integer range 0 to 3:=0;
begin
  if(rising_edge(clk_4MHz)) then
    if(count_1=0) then
      scl_buss<='0';
    elsif(count_1=1) then
      dcl_buss<='1';
    elsif(count_1=2) then
      scl_buss<='1';
    else
      dcl_buss<='0';
    end if;
    if(count_1=3) then
      count_1:=0;
    else
      count_1:=count_1 + 1;
    end if;
  end if;
end process;
```

PR 5.2 Program 5.2

In PR 5.3, the complete program for the generation of I2C bus-clock "scl" and data-clock "dcl" is given.

```vhdl
library ieee;
use ieee.std_logic_1164.all;

entity buss_clk_gen is
  port(clk_100MHz: in std_logic;
       scl, dcl: out std_logic);
end entity;

architecture logic_flow of buss_clk_gen is
  signal count: natural range 0 to 11;
  signal clk_4MHz: std_logic:='0';
  signal scl_buss, dcl_buss: std_logic:='0';
begin
  scl<=scl_buss; dcl<=dcl_buss;
  clk4MHz: process(clk_100MHz)
  begin
    if (rising_edge(clk_100MHz)) then
      count<=count + 1;
      if(count=12) then
        clk_4MHz<=not clk_4MHz;
        count<=0;
      end if;
    end if;
  end process;
```

```vhdl
  clk1MHz: process (clk_4MHz)
    variable count_1: integer range 0 to 3:=0;
  begin
    if(rising_edge(clk_4MHz)) then
      if(count_1=0) then
        scl_buss<='0';
      elsif(count_1=1) then
        dcl_buss<='1';
      elsif(count_1=2) then
        scl_buss<='1';
      else
        dcl_buss<='0';
      end if;
      if(count_1=3) then
        count_1:=0;
      else
        count_1:=count_1 + 1;
      end if;
    end if;
  end process;
end;
```

PR 5.3   Program 5.3

The program in PR 5.3 can be tested using the test-bench program given in PR 5.4.

```vhdl
library ieee;
use ieee.std_logic_1164.all;

entity buss_clk_gen_tb is
end;

architecture bench of buss_clk_gen_tb is

  component buss_clk_gen
    port(clk_100MHz: in std_logic;
         scl, dcl: out std_logic);
  end component;

  signal clk_100MHz: std_logic;
  signal scl, dcl: std_logic;
  constant clock_period: time:= 10 ns;
  signal stop_the_clock: boolean;

begin

  pm: buss_clk_gen port map(clk_100MHz => clk_100MHz,
                            scl         => scl,
                            dcl         => dcl );

  p1: process --stimulus
  begin
    wait for clock_period*10*100;
    stop_the_clock<=true;
    wait;
  end process;

  p2: process --clock generation
  begin
    while not stop_the_clock loop
      clk_100MHz<='0';
      wait for clock_period / 2;
      clk_100MHz<='1';
      wait for clock_period / 2;
    end loop;
    wait;
  end process;
end;
```

PR 5.4 Program 5.4

Now, let us consider the VHDL implementation of I2C communication protocol. We will explain the topic through an example.

Example 5.6 Implement the I2C protocol, where a master device sends only an 8-bit data to a slave device, in VHDL. Use all zero-bit sequences for 7-bit slave address. Use 1 MHz for I2C bus-clock frequency. The frame structure for this operation is depicted again in Fig. 5.25.

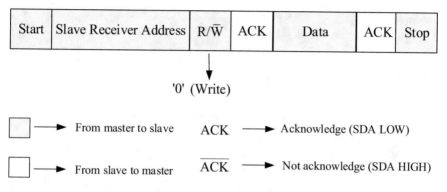

Fig. 5.25 Frame format for Example 5.6

Solution 5.6 We will use PR 5.3 in our program. We will use timed state machine for our implementation. The timed state diagram for this transmission is shown again in Fig. 5.26.

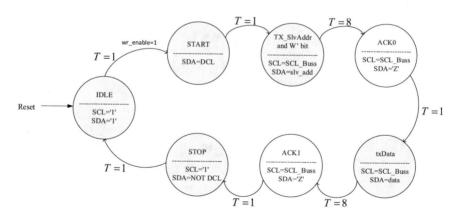

Fig. 5.26 State diagram for Fig. 5.25

Considering Fig. 5.26, we write the entity part of our program as in PR 5.5.

```vhdl
library ieee;
use ieee.std_logic_1164.all;
entity fsm_i2c is
  port(clk, rst, wr_enable: in std_logic;
       scl: out  std_logic;
       sda: inout std_logic);
end entity;

architecture logic_flow of fsm_i2c is
  type state is (st_idle, st0_start, st1_txSlaveAddress, st2_ack1, st3_txData, st4_ack2,
              st5_stop);
  signal present_state, next_state: state;
  signal scl_buss, dcl_buss: std_logic;
  constant data: std_logic_vector(7 downto 0):="11101100";
  constant slave_address_with_wrt_flg: std_logic_vector(7 downto 0):="11101100";
  constant max_length: natural:=8;
  signal data_index: natural range 0 to max_length -1;
  signal timer: natural range 0 to max_length;
  signal ack_bits: std_logic_vector(1 downto 0);
  signal count: natural range 0 to 11:=0;
  signal clk_4MHz: std_logic;
  begin
```

PR 5.5 Program 5.5

If we add the data-clock and bus-clock generator processes given in PR 5.5 in our program, we get PR 5.6.

```
library ieee;
use ieee.std_logic_1164.all;
entity fsm_i2c is
  port(clk_100MHz, rst, wr_enable: in std_logic;
       scl: out std_logic;
       sda: inout std_logic);
end entity;

architecture logic_flow of fsm_i2c is
  type state is (st_idle, st0_start, st1_txSlaveAddress, st2_ack0, st3_txData, st4_ack1,
               st5_stop);
  signal present_state, next_state: state;
  signal scl_buss, dcl_buss: std_logic;
  constant data: std_logic_vector(7 downto 0):="11101100";
  constant slave_address_with_wrt_flg: std_logic_vector(7 downto 0):="11101100";
  constant max_length: natural:=8;
  signal data_index: natural range 0 to max_length -1;
  signal timer: natural range 0 to max_length;
  signal ack_bits: std_logic_vector(1 downto 0);
  signal count: natural range 0 to 11:=0;
  signal clk_4MHz: std_logic;
begin
```

```
scl<=scl_buss;

clk4MHz: process(clk_100MHz)
begin

  if (rising_edge(clk_100MHz)) then
    count<= count + 1;
    if(count=12) then
      clk_4MHz<=not clk_4MHz;
      count<=0;
    end if;
  end if;

end process;
```

```
clk1MHz: process (clk_4MHz)
  variable count_1: integer range 0 to 3:=0;
begin
  if(rising_edge(clk_4MHz)) then
    count_1:= count_1 + 1;
    if(count_1=0) then
      scl_buss<='0';
    elsif(count_1=1) then
      dcl_buss<='1';
    elsif(count_1=2) then
      scl_buss<='1';
    else
      dcl_buss<='0';
    end if;
    if(count_1=3) then
      count_1:=0;
    else
      count_1:=count_1 + 1;
    end if;
  end if;
end process;
```

PR 5.6 Program 5.6

Now we need three more processes. One process, "p1", is used for the update of the present state. The second one, "p2", is used to read the data sent by the slave, i.e., for this example only acknowledgments bits, and the third one, "p3", is used for the determination of next states and output port values.

The first two processes "p1" and "p2" are given in PR 5.7.

```
p1: process(dcl_buss, rst)
begin
  if(rst ='1') then
    present_state<=st_idle;
    data_index<=0;
  elsif (dcl_buss 'event and dcl_buss ='1') then
    if(data_index=timer-1) then
      present_state<=next_state;
      data_index<=0;
    else
      data_index<=data_index +1;
    end if;
  end if;
end process;
```

```
p2: process(dcl_buss)
begin
  if(dcl_buss 'event and dcl_buss ='0') then
    if(present_state=st2_ack0) then
      ack_bits(0)<=sda;
    elsif( present_state=st4_ack1) then
      ack_bits(1)<=sda;
    end if;
  end if;
end process;
```

PR 5.7 Program 5.7

The third process "p3" is given in PR 5.8.

```
--- Circuit outputs and next states
p3: process(present_state, wr_enable, data_index, dcl_buss)
begin
  case present_state is
    when st_idle =>
      scl<='1';
      sda<='1';
      timer<=1;
      if(wr_enable='1') then
        next_state<= st0_start;
      else
        next_state<=st_idle;
      end if;

    when st0_start =>
      sda<=dcl_buss; scl<='1';
      timer<=1;
      next_state<= st1_txSlaveAddress;

    when st1_txSlaveAddress =>
      sda<=slave_address_with_wrt_flg(7- data_index);
      timer<=8; scl<=scl_buss;
      next_state<=st2_ack0;
```

```
    when st2_ack0=>
      sda<='Z'; scl<=scl_buss;
      timer<=1;
      next_state<=st3_txData;

    when st3_txData =>
      sda<=data(7-data_index);
      timer<=8; scl<=scl_buss;
      next_state<=st4_ack1;

    when st4_ack1=>
      sda<= 'Z'; scl<=scl_buss;
      timer<=1;
      next_state<=st5_stop;

    when st5_stop =>
      sda<=not dcl_buss;
      timer<=1; scl<='1';
      next_state<=st_idle;

  end case;
end process;
end logic_flow;
```

PR 5.8 Program 5.8

In PR 5.9, the structure of overall program is depicted.

```
library ieee;
use ieee.std_logic_1164.all;

entity fsm_i2c is
....
end entity;

architecture logic_flow of fsm_i2c is

 begin

clk4MHz: process(clk_100MHz)
....
end process;

clk1MHz: process (clk_4MHz)
....
end process;
```

```
p1: process(dcl_buss, rst)
 ....
  end process;

p2: process(dcl_buss)
 ....
  end process;

p3: process(present_state, wr_enable,...)
 ....
 end process;

end logic_flow;
```

PR 5.9   Program 5.9

If we combine all the program parts, we get the overall program as in PR 5.10.

```vhdl
library ieee;
use ieee.std_logic_1164.all;
entity fsm_i2c is
  port(clk_100MHz, rst, wr_enable: in std_logic;
       scl, dcl: out std_logic;
       sda: inout std_logic);
end entity;


architecture logic_flow of fsm_i2c is
  type state is (st_idle, st0_start, st1_txSlaveAddress, st2_ack0,
                 st3_txData, st4_ack1, st5_stop);
  signal present_state, next_state: state;
  signal scl_buss: std_logic:='0';
  signal dcl_buss: std_logic:='0';
  constant data: std_logic_vector(7 downto 0):="11101100";
  constant slave_address_with_wrt_flg:
                 std_logic_vector(7 downto 0):="11101100";
  constant max_length: natural:=8;
  signal data_index: natural range 0 to max_length -1;
  signal timer: natural range 0 to max_length;
  signal ack_bits: std_logic_vector(1 downto 0);
  signal count: natural range 0 to 11:=0;
  signal clk_4MHz: std_logic:='0';
begin

  dcl<=dcl_buss;
  clk4MHz: process(clk_100MHz)
  begin
    if(rising_edge(clk_100MHz)) then
      count<=count + 1;
      if(count=12) then
        clk_4MHz<=not clk_4MHz;
        count<=0;
      end if;
    end if;
  end process;
```

```vhdl
clk1MHz: process (clk_4MHz)
  variable count_1: integer range 0 to 3:=0;
begin
  if(rising_edge(clk_4MHz)) then
    if(count_1=0) then
      scl_buss<='0';
    elsif(count_1=1) then
      dcl_buss<='1';
    elsif(count_1=2) then
      scl_buss<='1';
    else
      dcl_buss<='0';
    end if;
    if(count_1=3) then
      count_1:=0;
    else
      count_1:=count_1 + 1;
    end if;
  end if;
end process;
p1: process(dcl_buss, rst)
begin
  if(rst ='1') then
    present_state<=st_idle;
    data_index<=0;
  elsif(dcl_buss'event and dcl_buss='1') then
    if(data_index=timer-1) then
      present_state<=next_state;
      data_index<=0;
    else
      data_index<=data_index +1;
    end if;
  end if;
end process;

p2: process(dcl_buss)
begin
  if(dcl_buss'event and dcl_buss='1') then
    if(present_state=st2_ack0) then
      ack_bits(0)<=sda;
    elsif( present_state=st4_ack1) then
      ack_bits(1)<=sda;
    end if;
  end if;
end process;
```

PR 5.10 Program 5.10

```vhdl
--- Circuit outputs and next states
p3: process(present_state, wr_enable, data_index, dcl_buss)
begin
  case present_state is
    when st_idle =>
      scl<='1';
      sda<='1';
      timer<=1;
      if(wr_enable='1') then
        next_state<=st0_start;
      else
        next_state<=st_idle;
      end if;

    when st0_start =>
      sda<=dcl_buss;
      timer<=1;
      next_state<=st1_txSlaveAddress;
      scl<='1';
    when st1_txSlaveAddress =>
      sda<=slave_address_with_wrt_flg(7-data_index);
      timer<=8;
      next_state<=st2_ack0;
      scl<=scl_buss;
    when st2_ack0=>
      sda<='Z';
      timer<=1;
      next_state<=st3_txData;
      scl<=scl_buss;
    when st3_txData =>
      sda<=data(7-data_index);
      timer<=8;
      next_state<=st4_ack1;
      scl<=scl_buss;
    when st4_ack1=>
      sda<='Z';
      timer<=1;
      next_state<=st5_stop;
      scl<=scl_buss;
    when st5_stop =>
      sda<=not dcl_buss;
      timer<=1;
      next_state<=st_idle;
      scl<='1';
  end case;
end process;

end logic_flow;
```

PR 5.10   (continued)

The program in PR 5.10 can be tested using the test-bench program in PR 5.11.

```vhdl
library ieee;
use ieee.std_logic_1164.all;

entity fsm_i2c_tb is
end;

architecture bench of fsm_i2c_tb is

  component fsm_i2c
   port(clk_100MHz, rst, wr_enable:  in std_logic;
        scl, dcl: out  std_logic;
        sda: inout std_logic);
  end component;

  signal clk_100MHz, rst, wr_enable: std_logic;
  signal scl, dcl: std_logic;
  signal sda: std_logic;
  constant clock_period: time:= 10 ns;
  signal stop_the_clock: boolean;

begin

  pm: fsm_i2c port map(clk_100MHz => clk_100MHz,
                       rst        => rst,
                       wr_enable  => wr_enable,
                       scl        => scl,
                       dcl        => dcl,
                       sda        => sda );
  p1: process  --stimulus
  begin
   rst<='1'; rst<='0';
   wait for clock_period*100;
   wr_enable<='0'; wr_enable<='1';

   wait for clock_period*22*100;
   wr_enable<='0';
   stop_the_clock<=true;
   wait;
  end process;

  p2: process --clock generation
  begin
   while not stop_the_clock loop
    clk_100MHz<= '0';
    wait for clock_period / 2;
    clk_100MHz<= '1';
    wait for clock_period / 2;
   end loop;
   wait;
  end process;
end;
```

PR 5.11 Program 5.11

Exercise The frame structure for the data transmitted from master to slave, i.e., write operation, is shown in Fig. 5.27.

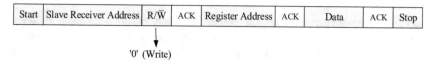

Start	Slave Receiver Address	R/W̄	ACK	Register Address	ACK	Data	ACK	Stop

'0' (Write)

Fig. 5.27 Frame format for exercise

The state diagram for this write operation can be drawn as in Fig. 5.28. Implement the I2C communication protocol in VHDL. Use dummy values for slave, register addresses, and data value.

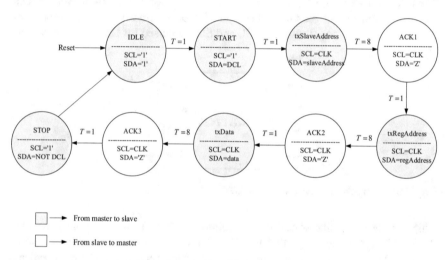

Fig. 5.28 State diagram for Fig. 5.27

Exercise I2C communication frame format for the master to read data from the slave is depicted in Fig. 5.29 where a single byte is read from the slave. Draw the state diagram for this I2C communication scheme and implement it in VHDL. Use dummy values for slave address and data value.

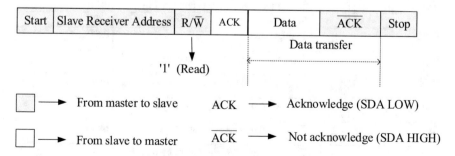

Start	Slave Receiver Address	R/W̄	ACK	Data	ACK̄	Stop

'1' (Read)

Data transfer

☐ ⟶ From master to slave   ACK ⟶ Acknowledge (SDA LOW)

☐ ⟶ From slave to master   ACK̄ ⟶ Not acknowledge (SDA HIGH)

Fig. 5.29 I2C communication frame format for the master to read data from the slave

5.7   VHDL Implementation of FPGA and ADT7420 I2C Interfacing

ADT7420 is a high precision digital temperature sensor used in the industry. It consists of a band-gap reference, a temperature sensor, and a 16-bit ADC for the digitization of the temperature with 0.0078 °C resolution. The resolution of the ADC, by default, is set to 13 bits (0.0625 °C). The resolution of ADC can be programmed by users, and it can be changed through the serial interface.

The operation limits of ADT7420 range from −40 °C to +150 °C. The black-box representation of ADT7420 is depicted in Fig. 5.30 where A0 and A1 pins are used for address selection. Using A0 and A1, four possible I2C addresses can be selected. The CT, and INT pins, which can operate in comparator and interrupt event modes, are open-drain output pins, and these pins become active when the temperature exceeds a programmed threshold. SCL and SDA are used for I2C communication.

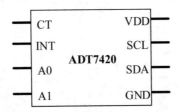

Fig. 5.30 The black-box representation of ADT7420

Chip configuration like selection of resolution, setting hysteresis point, checking device ID can be done via I2C interface. In Table 5.1, register address map of ADT7420 is shown. As it is seen from the table, some registers can be both read and written (indicated by R/$\overline{W}$), whereas some of them can only be read (indicated by R). Register addresses are represented with 8 bits.

Table 5.1 Register table of ADT7420

Register Address	Type	Description	Power-On Default
0x00	R	Temperature value most significant byte	0x00
0x01	R	Temperature value least significant byte	0x00
0x02	R	Status	0x00
0x03	R/$\overline{W}$	Configuration	0x00
0x04	R/$\overline{W}$	T_{HIGH} setpoint most significant byte	0x20
0x05	R/$\overline{W}$	T_{HIGH} setpoint least significant byte	0x00
0x06	R/$\overline{W}$	T_{LOW} setpoint most significant byte	0x05
0x07	R/$\overline{W}$	T_{LOW} setpoint least significant byte	0x00
0x08	R/$\overline{W}$	T_{CRIT} setpoint most significant byte	0x49

(continued)

Table 5.1   (continued)

Register Address	Type	Description	Power-On Default
0x09	R/$\overline{\text{W}}$	T_{CRIT} setpoint least significant byte	0x80
0x0A	R/$\overline{\text{W}}$	T_{HYST} setpoint	0x05
0x0B	R	ID	0xCB
0x0C	R/$\overline{\text{W}}$	Reserved	0xXX
0x0D	R/$\overline{\text{W}}$	Reserved	0xXX
0x2E	R/$\overline{\text{W}}$	Reserved	0xXX
0x2F	R/$\overline{\text{W}}$	Software reset	0xXX

5.7.1   VHDL Implementation of I2C Communication Between FPGA and ADT7420

The circuit used for FPGA and ADT7420 interfacing is depicted in Fig. 5.31. Electronic devices supporting I2C protocol have slave addresses. ADT7420 IC has 7-bit slave address in which 5 bits ("10010") are fixed and the remaining 2 bits are adjustable, and adjustable bits are determined using A0 and A1. It is seen in Fig. 5.31 that A0 and A1 are connected to the power supply, i.e., A0 and A1 have logic-1. In this case, the slave address of ADT7420 becomes "1001011". Furthermore, both SCL and SDA are of open-drain ports, for this reason, pull-up resistors are connected.

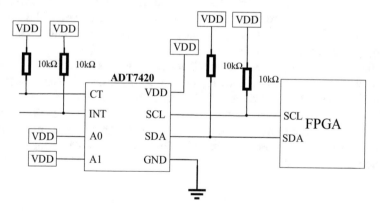

Fig. 5.31   ADT7420 and FPGA interfacing

5.7.1.1 ID Register of ADT7420

It is shown in Table 5.1 that ADT7420 has 8-bit manufacturer ID which is stored in 8-bit register whose address is 0x0B, and the value of manufacturer ID is 0xCB, i.e., 8-bit register contains 0xCB.

5.7.1.2 VHDL Implementation

Now we will consider the VHDL implementation I2C serial communication between ADT7420 and FPGA via an example.

Example 5.7 Read the manufacture ID of ADT7420 via I2C interface. Show the 8-bit ID value on your FPGA board's LEDs. Use the circuit shown in Fig. 5.31 while writing your VHDL program.

Solution 5.7 The complete frame format to read the manufacturer ID of ADT7420 via I2C communication is depicted in Fig. 5.32.

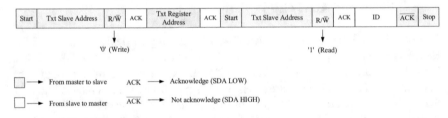

Fig. 5.32 Frame format for Example 5.7

It is seen from Fig. 5.32 that, first the slave address with $\overline{\text{W}}$, i.e., "10010110", is transmitted from master to slave, and ACK is received by the master. Next, register address, 0x0B, where manufacture ID value is available, is transmitted. This is followed by ACK signal received by master.

In the next stage of the transmission, first a start signal is transmitted. This is followed by the transmission of slave address and read bit, i.e., R. Then, master gets the ACK followed by 8-bit device ID. Master ends the communication sending a negative acknowledgment signal followed by a stop signal. In Fig. 5.32, communication logic is illustrated using frames; however, no timing information is provided. The transmission frame format depicted in Fig. 5.32 is illustrated using the timing waveforms in Fig. 5.33.

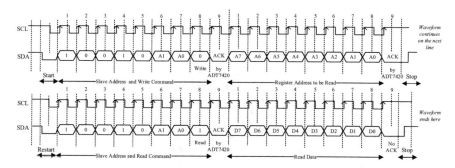

Fig. 5.33 Transmission timing waveforms for Fig. 5.32

State diagram corresponding to the transmission waveforms of Fig. 5.33 can be drawn as in Fig. 5.34.

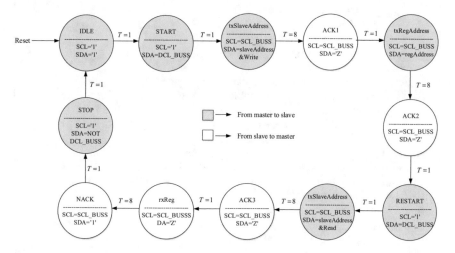

Fig. 5.34 State diagram for Fig. 5.33

The operations performed by the master unit in the state diagram of Fig. 5.34 can be written in sequel as

1. Send START signal.
2. Send slave address concatenated with the WRITE flag.
3. Get ACK.
4. Send the register address to be read.
5. Get ACK.
6. Send START signal again.
7. Send slave address concatenated with the READ flag.
8. Get ACK.

9. Read the 8-bit register data.
10. Send Negative ACK signal.
11. Send STOP signal.

In PR 5.12, entity part is written and signal, constant and variable object declarations are made in the declarative part of the architecture unit. Slave address concatenated with WRITE and READ flags are defined as constant objects. Moreover, slave register address (0x0b) that contains the device manufacture ID is also defined as constant object. The other objects defined are used in auxiliary clock signal generation, clock division operation, holding register indices, and holding ACK information.

```vhdl
library ieee;
use ieee.std_logic_1164.all;
entity fsm_i2c_exp1 is
  port(clk_100MHz, rst: in std_logic;
       scl: out std_logic;
       sda: inout std_logic;
       led: out std_logic_vector(10 downto 0)
  );
  end entity;
architecture logic_flow of fsm_i2c_exp1 is
  type state is (st_idle, st0_start, st1_txSlaveAddress, st2_ack1, st3_txAddress,
                 st4_ack2, st5_restart, st6_txSlaveAddress, st7_ack3,
                 st8_rd_data, st9_nack, st10_stop);
  signal DataOut: std_logic_vector(7 downto 0);
  constant Address_tobe_Read: std_logic_vector(7 downto 0):=x"0b";
  constant slave_address_with_rd_flg: std_logic_vector(7 downto 0):=x"97";
  constant slave_address_with_wrt_flg: std_logic_vector(7 downto 0):=x"96";
  signal scl_buss: std_logic:='0';
  signal present_state,next_state:state ;
  signal dcl_buss: std_logic:='0';
  constant max_length: integer:=8;
  shared variable data_index: integer range 0 to max_length -1;
  signal timer: integer range 0 to max_length;
  signal clk_400KHz: std_logic:='0';
  signal ack_bits: std_logic_vector(2 downto 0);
  signal count: integer range 0 to 250:=0;
  signal sda_signal, scl_signal: std_logic;
  signal rd_flag: std_logic:='0';
begin
```

PR 5.12 Program 5.12

ADT7420 supports up to 400 KHz I2C communication speed. We choose 100 KHz transmission speed. 100 KHz data and bus-clock signals are generated from 100 MHz FPGA clock frequency using clock dividers in PR 5.13.

```
clk400KHz: process(clk_100MHz)
begin
 if(rst='1') then
   clk_400KHz<='0';
   count<=0;
 elsif(rising_edge(clk_100MHz)) then
   if(count=124) then
     clk_400KHz<=not clk_400KHz;
     count<=0;
   else
     count<=count + 1;
   end if;
 end if;
end process;

clk_100KHz: process (clk_400KHz)
 variable count_1: integer range 0 to 3:=0;
begin
 if(rst='1') then
   scl_buss<='1';
```

```
   dcl_buss<='1';
   count_1:=0;
 elsif(rising_edge(clk_400KHz)) then
   if(count_1=0) then
     scl_buss<='0';
   elsif(count_1=1) then
     dcl_buss<='1';
   elsif(count_1=2) then
     scl_buss<='1';
   else
     dcl_buss<='0';
   end if;
     if(count_1=3) then
       count_1:=0;
     else
       count_1:=count_1 + 1;
     end if;
 end if;
end process;
```

PR 5.13 Program 5.13

The processes "p1" and "p2" written in PR 5.14 are used to update present state, and to receive the acknowledgment signals sent by the slave.

```
p1: process(dcl_buss, rst)
  begin
   if (rst ='1') then
     present_state<=st_idle;
     data_index:=0;
   elsif (dcl_buss 'event and dcl_buss ='1') then
     if(data_index=timer-1) then
       present_state<=next_state;
       data_index:=0;
     else
       data_index:=data_index +1;
     end if;
   end if;
  end process;
led(7 downto 0)<= DataOut(7 downto 0);
led(10 downto 8)<= ack_bits;
```

```
p2: process(dcl_buss, rst)
begin
  if(dcl_buss'event and dcl_buss='0') then
   if(present_state=st2_ack1) then
     ack_bits(0)<=sda;
   elsif( present_state=st4_ack2) then
     ack_bits(1)<=sda;
   elsif (present_state=st7_ack3) then
     ack_bits(2)<=sda;
   elsif (present_state=st8_rd_data) then
     DataOut(7-data_index) <= sda;
   end if;
  end if;
end process;
```

PR 5.14 Program 5.14

The third process "p3" is written for the determination of next states and port outputs. The generation of start signal is illustrated on the right of PR 5.15 and its VHDL implementation is provided on the left of PR 5.15.

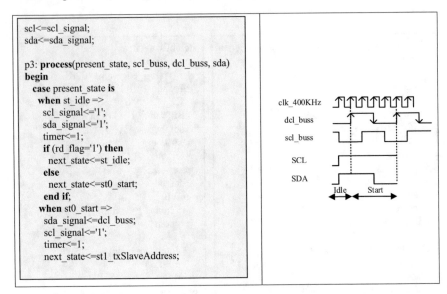

```
scl<=scl_signal;
sda<=sda_signal;

p3: process(present_state, scl_buss, dcl_buss, sda)
begin
  case present_state is
    when st_idle =>
      scl_signal<='1';
      sda_signal<='1';
      timer<=1;
      if (rd_flag='1') then
        next_state<=st_idle;
      else
        next_state<=st0_start;
      end if;
    when st0_start =>
      sda_signal<=dcl_buss;
      scl_signal<='1';
      timer<=1;
      next_state<=st1_txSlaveAddress;
```

PR 5.15 Program 5.15

In PR 5.16, determination of serial data, serial clock, timer, and next state values for the present state st1_txSlaveAddress are illustrated and written.

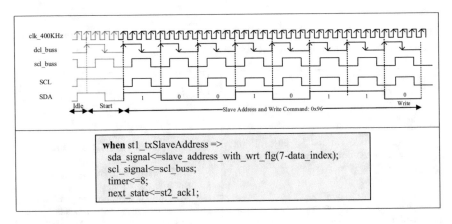

```
when st1_txSlaveAddress =>
  sda_signal<=slave_address_with_wrt_flg(7-data_index);
  scl_signal<=scl_buss;
  timer<=8;
  next_state<=st2_ack1;
```

PR 5.16 Program 5.16

After the transmission of slave address, acknowledgment is got from slave device. In receiver mode, the master releases the data bus and puts the data line at high impedance. In PR 5.17, determination of serial data, serial clock, timer, and next state values for the present state st1_txSlaveAddress are illustrated and written. Data reception is performed in PR 5.14 at the falling edge of the clock pulse.

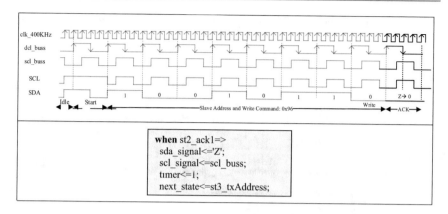

PR 5.17 Program 5.17

In PR 5.18, the address of register to be read is transmitted to the slave by the master. Transmission takes eight clock cycles, and the most significant bit of the address data is sent first. After reception of the acknowledgment bit, start signal is retransmitted.

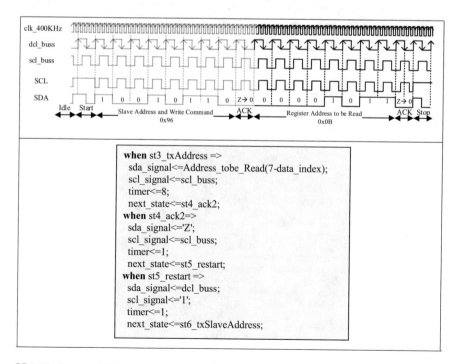

PR 5.18 Program 5.18

After the retransmission of START signal, slave address concatenated with READ flag is sent and ACK signal is received. These events are implemented in PR 5.19 and PR 5.14.

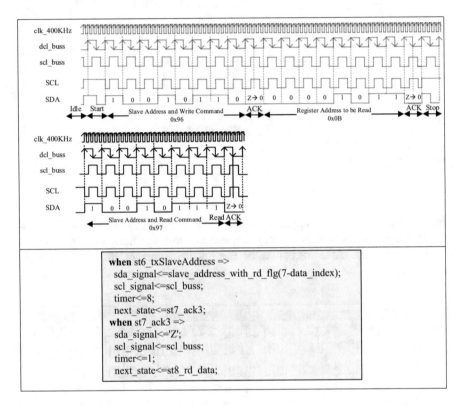

```
when st6_txSlaveAddress =>
  sda_signal<=slave_address_with_rd_flg(7-data_index);
  scl_signal<=scl_buss;
  timer<=8;
  next_state<=st7_ack3;
when st7_ack3 =>
  sda_signal<='Z';
  scl_signal<=scl_buss;
  timer<=1;
  next_state<=st8_rd_data;
```

PR 5.19　Program 5.19

Reading of device ID followed by the transmission of negative acknowledgment signal and STOP signal are implemented in PR 5.20, and in "p2" of PR 5.14.

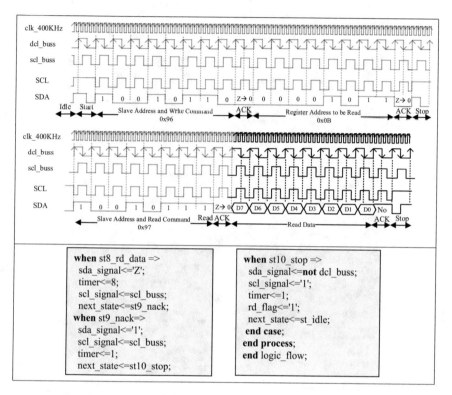

```
when st8_rd_data =>                      when st10_stop =>
  sda_signal<='Z';                         sda_signal<=not dcl_buss;
  timer<=8;                                 scl_signal<='1';
  scl_signal<=scl_buss;                     timer<=1;
  next_state<=st9_nack;                     rd_flag<='1';
when st9_nack=>                             next_state<=st_idle;
  sda_signal<='1';                        end case;
  scl_signal<=scl_buss;                   end process;
  timer<=1;                               end logic_flow;
  next_state<=st10_stop;
```

PR 5.20   Program 5.20

Combining all program parts, we get the overall VHDL code as in PR 5.21.

```vhdl
library ieee;
use ieee.std_logic_1164.all;
entity fsm_i2c_exp1 is
 port(clk_100MHz, rst: in std_logic;
     scl: out std_logic;
     sda: inout std_logic;
     led: out std_logic_vector(10 downto 0)
);
end entity
architecture logic_flow of fsm_i2c_exp1 is
 type state is (st_idle, st0_start, st1_txSlaveAddress,
        st2_ack1, st3_txAddress, st4_ack2, st5_restart,
        st6_txSlaveAddress, st7_ack3, st8_rd_data,
        st9_nack, st10_stop);
 signal DataOut: std_logic_vector(7 downto 0);
 constant Address_tobe_Read:
           std_logic_vector(7 downto 0):=x"0b";
 constant slave_address_with_rd_flg:
           std_logic_vector(7 downto 0):=x"97";
 constant slave_address_with_wrt_flg:
           std_logic_vector(7 downto 0):=x"96";
 signal scl_buss: std_logic:='0';
 signal dcl_buss: std_logic:='0';
 constant max_length: integer:=8;
 signal present_state,next_state  :state ;
 shared variable data_index: integer range 0 to
                     max_length -1;
 signal timer: integer range 0 to max_length;
 signal clk_400KHz: std_logic:='0';
 signal ack_bits: std_logic_vector(2 downto 0);
 signal count: integer range 0 to 250:=0;
 signal sda_signal, scl_signal: std_logic;
 signal rd_flag: std_logic:='0';
begin

clk400KHz: process(clk_100MHz)
begin
 if(rst='1') then
   clk_400KHz<='0';
   count<=0;
   count<=count + 1;
   end if;
  end if;
 end process;

 clk_100KHz: process (clk_400KHz)
  variable count_1: integer range 0 to 3:=0;
 begin
  if(rst='1') then
   scl_buss<='1'; dcl_buss<='1';
   count_1:=0;
  elsif(rising_edge(clk_400KHz)) then
   if(count_1=0) then
    scl_buss<='0';
   elsif(count_1=1) then
    dcl_buss<='1';
   elsif(count_1=2) then
    scl_buss<='1';
   else
    dcl_buss<='0';
   end if;
   if(count_1=3) then
    count_1:=0;
   else
    count_1:=count_1 + 1;
   end if;
  end if;
 end process;
p1: process(dcl_buss, rst)
 begin
  if (rst ='1') then
   present_state<=st_idle;
   data_index:=0;
  elsif (dcl_buss 'event and dcl_buss ='1') then
   if(data_index=timer-1) then
    present_state<=next_state;
    data_index:=0;
   else
    data_index:=data_index +1;
   end if;
  end if;
 end process;

led(7 downto 0)<=DataOut(7 downto 0);
led(10 downto 8)<=ack_bits;

p2: process(dcl_buss, rst)
begin
 if(dcl_buss'event and dcl_buss='0') then
  if(present_state=st2_ack1) then
   ack_bits(0)<=sda;
  elsif( present_state=st4_ack2) then
   ack_bits(1)<=sda;
  elsif (present_state=st7_ack3) then
   ack_bits(2)<=sda;
  elsif (present_state=st8_rd_data) then
   DataOut(7-data_index)<=sda;
  end if;
 end if;
end process;
scl<=scl_signal;
sda<=sda_signal;
p3: process(present_state, scl_buss, dcl_buss, sda)
begin
  case present_state is
   when st_idle =>
    scl_signal<='1';
    sda_signal<='1';
    timer<=1;
    if(rd_flag='1') then
     next_state<=st_idle;
    else
     next_state<=st0_start;
    end if;
   when st0_start =>
    sda_signal<=dcl_buss;
    scl_signal<='1';
    timer<=1;
    next_state<=st1_txSlaveAddress;
   when st1_txSlaveAddress =>
    sda_signal<=slave_address_with_wrt_flg
(7-data_index);
    scl_signal<=scl_buss;
    timer<=8;
    next_state<=st2_ack1;
   when st2_ack1=>
    sda_signal<='Z';
    scl_signal<=scl_buss;

    timer<=1;
    next_state<=st3_txAddress;
   when st3_txAddress =>
    sda_signal<=Address_tobe_Read(7-data_index);
    scl_signal<=scl_buss;
    timer<=8;
    next_state<=st4_ack2;
   when st4_ack2 =>
    sda_signal<='Z';
    scl_signal<=scl_buss;
    timer<=1;
    next_state<=st5_restart;
   when st5_restart =>
    sda_signal<=dcl_buss;
    scl_signal<='1';
    timer<=1;
    next_state<=st6_txSlaveAddress;
   when st6_txSlaveAddress =>
    sda_signal<=
       slave_address_with_rd_flg(7-data_index);
    scl_signal<=scl_buss;
    timer<=8;
    next_state<=st7_ack3;
   when st7_ack3 =>
    sda_signal<='Z';
    scl_signal<=scl_buss;
    timer<=1;
    next_state<=st8_rd_data;
   when st8_rd_data =>
    sda_signal<='Z';
    scl_signal<=scl_buss;
    timer<=8;next_state<=st9_nack;
   when st9_nack=>
    sda_signal<='1';
    scl_signal<=scl_buss;
    timer<=1; next_state<=st10_stop;
   when st10_stop =>
    sda_signal<=not dcl_buss;
    scl_signal<='1';
    timer<=1;
    rd_flag<='1';
    next_state<=st_idle;
   end case;end process;
   end logic_flow;
```

PR 5.21 Program 5.21

Example 5.8 In this example, we implement the write and the read operations for ADT7420. Write the value 0xC1 to the register of ADT7420 whose address is 0x04. After write operation, read the content of register whose address is 0x04.

Solution 5.8 The timing waveforms of the I2C communication in write operation for ADT7420 is depicted in Fig. 5.35. Writing to a register of ADT7420 consists of three main steps as illustrated in Fig. 5.35. In the first step, slave address concatenated with WRITE flag is transmitted. In the second step, register address is transmitted. In the last step, data to be written to the register, starting from the most significant bit, is transmitted. Between every two transmissions, acknowledgment, sent by slave device, is received by the master (FPGA).

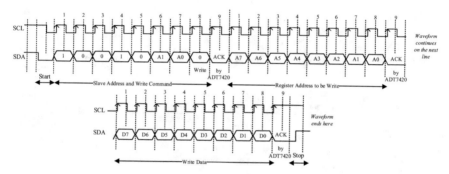

Fig. 5.35 Transmission timing waveforms for Example 5.8

The state diagram corresponding to the I2C waveforms depicted in Fig. 5.35 can be drawn as in Fig. 5.36.

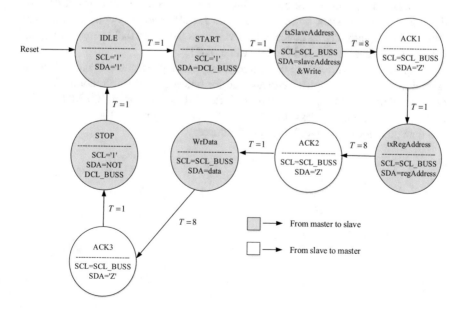

Fig. 5.36 State diagram for Fig. 5.35

In our previous example, we implemented the read operation for ADT7420. The state diagram of the read operation is given in Fig. 5.34. We can combine state diagram used for read and write operations. When Figs. 5.34 and 5.36 are inspected, it is seen that the states "Idle", "Start", "txSlaveAddress", and "ACK1" are used in both reading and writing operations. Considering this issue, we can combine both state diagrams as in Fig. 5.37 where a new control signal "write_done" is introduced to enable the read operation.

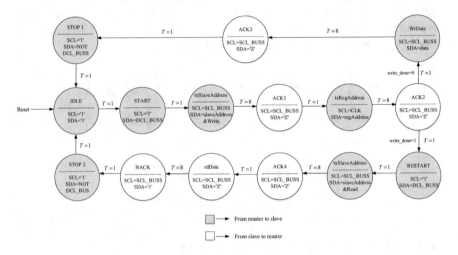

Fig. 5.37 Combined state diagram

In PR 5.22, entity part is written and signal, constant, and variable object declarations are made. Constant objects are defined for slave address concatenated with write and read flags having concatenated values of 0x96 and 0x97 respectively, and for register address 0x04, and for data value 0xC1.

Other object definitions are used in auxiliary clock generation, in clock division operation, for holding register indices, and for holding ACK information. The process "clk400KHz" is used to obtain a clock of 400 KHz from 100 MHz FPGA clock, and the second process "clk_100KHz" is used to generate 100 KHz serial bus and data bus-clocks from 400 KHz clock source.

```vhdl
library ieee;
use ieee.std_logic_1164.all;
entity fsm_i2c_exp2 is
  port(clk_100MHz, rst: in std_logic;
       scl: out std_logic;
       sda: inout std_logic;
       led: out std_logic_vector(11 downto 0) );
end;
architecture logic_flow of fsm_i2c_exp2 is
  type state is (st_idle, st0_start, st1_txSlaveAddress, st2_ack1, st3_txAddress,
                 st4_ack2, st5_wr_data, st6_ack3, st7_stop1, st8_restart,
                 st9_txSlaveAddress, st10_ack4, st11_rd_data, st12_nack, st13_stop2);
  signal present_state, next_state: state;

  signal DataOut:  std_logic_vector(7 downto 0);
  constant data_to_write:  std_logic_vector(7 downto 0):=x"c1";
  constant Address_tobe_Read:  std_logic_vector(7 downto 0):=x"04";
  constant slave_address_with_rd_flg: std_logic_vector(7 downto 0):=x"97";
  constant slave_address_with_wrt_flg: std_logic_vector(7 downto 0):=x"96";
  signal scl_buss, dcl_buss: std_logic:='0';
  constant max_length: integer:=8;
  shared variable data_index: integer range 0 to max_length -1;
  signal timer: integer range 0 to max_length;
  signal ack_bits: std_logic_vector(3 downto 0);
  signal count: integer range 0 to 250:=0;
  signal clk_400KHz: std_logic:='0';
  signal sda_signal, scl_signal : std_logic;
  signal rd_flag: std_logic:='0';
  signal write_done: std_logic:='0';
begin
```

```vhdl
clk400KHz: process(clk_100MHz)
begin
if(rst='1') then
  clk_400KHz <='0';
  count<=0;
  elsif(rising_edge(clk_100MHz)) then
    if(count=124) then
      clk_400KHz<=not clk_400KHz;
      count<=0;
    else
      count<=count + 1;
    end if;
  end if;
end process;
clk_100KHz: process (clk_400KHz)
  variable count_1: integer range 0 to 3:=0;
begin
  if(rst='1') then
    scl_buss<='1';
```

```vhdl
  dcl_buss<='1';
  count_1:=0;
  elsif(rising_edge(clk_400KHz)) then
    if(count_1=0) then
      scl_buss<='0';
    elsif(count_1=1) then
      dcl_buss<='1';
    elsif(count_1=2) then
      scl_buss<='1';
    else
      dcl_buss<='0';
    end if;
    if(count_1=3) then
      count_1:=0;
    else
      count_1:=count_1 + 1;
    end if;
  end if;
end process;
```

PR 5.22   Program 5.22

The process "p1" written in PR 5.23 is used to update the present state, and the process "p2" written in PR 5.23 is used for the retrieval of acknowledgment and data.

```
p1: process(dcl_buss, rst)
begin
  if(rst ='1') then
    present_state<=st_idle;
    data_index<=0;
  elsif (dcl_buss 'event and dcl_buss ='1') then
    if(data_index=timer-1) then
      present_state<=next_state;
      data_index<=0;
    else
      data_index<=data_index +1;
    end if;
  end if;
end process;
led(7 downto 0)<= DataOut(7 downto 0);
led(11 downto 8)<= ack_bits;
```

```
p2: process(dcl_buss, rst)
begin
  if(dcl_buss'event and dcl_buss='0') then
    if(present_state=st2_ack1) then
      ack_bits(0)<=sda;
    elsif( present_state=st4_ack2) then
      ack_bits(1)<=sda;
    elsif (present_state=st6_ack3) then
      ack_bits(2)<=sda;
    elsif (present_state=st10_ack4) then
      ack_bits(3)<=sda;
    elsif (present_state=st11_rd_data) then
      DataOut(7-data_index) <= sda;
    end if;
  end if;
end process;
```

PR 5.23   Program 5.23

The third process "p3" is used to determine the next states and port outputs. In PR 5.24, the implementation of the write operation is made.

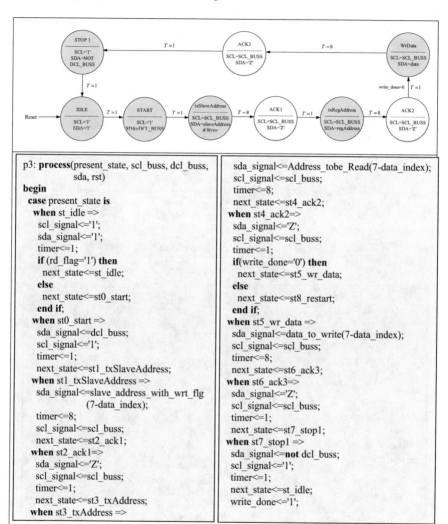

```
p3: process(present_state, scl_buss, dcl_buss,
            sda, rst)
begin
 case present_state is
  when st_idle =>
   scl_signal<='1';
   sda_signal<='1';
   timer<=1;
   if (rd_flag='1') then
    next_state<=st_idle;
   else
    next_state<=st0_start;
   end if;
  when st0_start =>
   sda_signal<=dcl_buss;
   scl_signal<='1';
   timer<=1;
   next_state<=st1_txSlaveAddress;
  when st1_txSlaveAddress =>
   sda_signal<=slave_address_with_wrt_flg
              (7-data_index);
   timer<=8;
   scl_signal<=scl_buss;
   next_state<=st2_ack1;
  when st2_ack1=>
   sda_signal<='Z';
   scl_signal<=scl_buss;
   timer<=1;
   next_state<=st3_txAddress;
  when st3_txAddress =>
```

```
   sda_signal<=Address_tobe_Read(7-data_index);
   scl_signal<=scl_buss;
   timer<=8;
   next_state<=st4_ack2;
  when st4_ack2=>
   sda_signal<='Z';
   scl_signal<=scl_buss;
   timer<=1;
   if(write_done='0') then
    next_state<=st5_wr_data;
   else
    next_state<=st8_restart;
   end if;
  when st5_wr_data =>
   sda_signal<=data_to_write(7-data_index);
   scl_signal<=scl_buss;
   timer<=8;
   next_state<=st6_ack3;
  when st6_ack3=>
   sda_signal<='Z';
   scl_signal<=scl_buss;
   timer<=1;
   next_state<=st7_stop1;
  when st7_stop1 =>
   sda_signal<=not dcl_buss;
   scl_signal<='1';
   timer<=1;
   next_state<=st_idle;
   write_done<='1';
```

PR 5.24 Program 5.24

The read operation is performed using the first five states and the remaining states are shown in PR 5.25. The flag "write_done" is used to indicate whether write operation is completed or not and it is checked in state "st4_ack2" in PR 5.24 to choose the upper or lower part of the state diagram of Fig. 5.30. In PR 5.24, "write_done" is assigned to logic-1 at the end of the state in "st7_stop1".

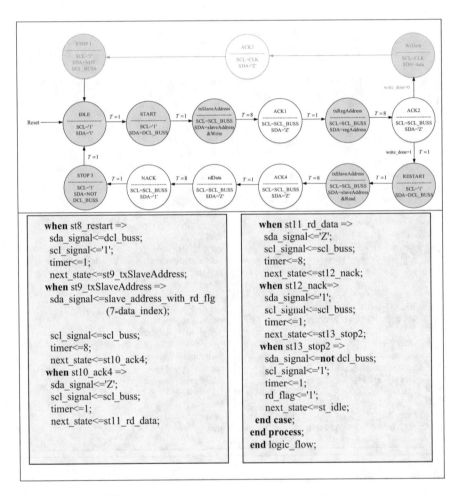

```
when st8_restart =>                          when st11_rd_data =>
  sda_signal<=dcl_buss;                        sda_signal<='Z';
  scl_signal<='1';                             scl_signal<=scl_buss;
  timer<=1;                                    timer<=8;
  next_state<=st9_txSlaveAddress;              next_state<=st12_nack;
when st9_txSlaveAddress =>                    when st12_nack=>
  sda_signal<=slave_address_with_rd_flg          sda_signal<='1';
            (7-data_index);                    scl_signal<=scl_buss;
                                               timer<=1;
  scl_signal<=scl_buss;                        next_state<=st13_stop2;
  timer<=8;                                  when st13_stop2 =>
  next_state<=st10_ack4;                       sda_signal<=not dcl_buss;
when st10_ack4 =>                              scl_signal<='1';
  sda_signal<='Z';                             timer<=1;
  scl_signal<=scl_buss;                        rd_flag<='1';
  timer<=1;                                    next_state<=st_idle;
  next_state<=st11_rd_data;                  end case;
                                           end process;
                                           end logic_flow;
```

PR 5.25 Program 5.25

Problems

1. Draw the I2C communication waveforms for the transmission of the single data byte, "10110101". For the slave address, use "0000000".
2. Master performs an immediate read slave operation. Only a single byte is read from slave. First draw the frame structure for this operation, then draw the state diagram of I2C transmission scheme. Use dummy values for slave address and data value.
3. Master performs a write operation which is followed by an immediate read operation. In both operations, a single byte is written and a single byte is read. Draw the frame structure for this I2C communication, draw the state diagram and implement it in VHDL. Use dummy values for slave address and data values.
4. Draw the frame format of I2C communication protocol where first write operation is performed, second an immediate read operation is performed, and in sequel another write operation is performed. Only single bytes are written and read.
5. Draw the state diagram of Fig. 5.22.
6. State the differences between SPI and I2C serial communications.
7. State the difference between RS232 and I2C serial communications.
8. Write a process to obtain 2 KHz bus and data clocks from 100 MHz FPGA's clock.

Video Graphic Array (VGA) and HDMI Interfacing

6

VGA is an interface introduced by IBM company in 1987 for the transmission of display data from computers to monitors. In this standard, only display data transmission is performed. Audio data transmission is not supported by VGA standard. A VGA connector contains both analog and digital data transmission lines. High-Definition Multimedia Interface (HDMI) is developed by a number of companies in the year 2002. It is used for transmitting uncompressed video data and compressed or uncompressed digital audio data. In this chapter, we will first provide brief information about VGA and HDMI interface, and then explain VHDL implementation of VGA and HDMI interface. VHDL codes are written in a stepwise manner for the clear understanding of the subject.

6.1   Video Graphic Array (VGA)

Video graphic array (VGA) interfacing is used for connecting computers to monitors. The VGA interfacing circuitry differs at computer and monitor sides. At the computer side, VGA interfacing circuitry has a graphic controller, on the other hand, on the monitor side it has a display controller as shown in Fig. 6.1.

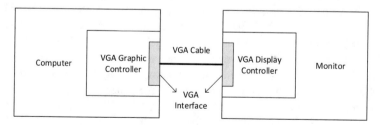

Fig. 6.1 VGA interface between computer and monitor

6.1.1 Graphic Controller

Graphic controller is an electronic system used to provide synchronization signals which are used to control the formation of images on monitor's screen. Besides, graphic controller also provides color signals which are used to give the colors of pixels. The signals between graphic controller and monitor are depicted in Fig. 6.2, and these signals are carried by VGA cables between the computer and monitor.

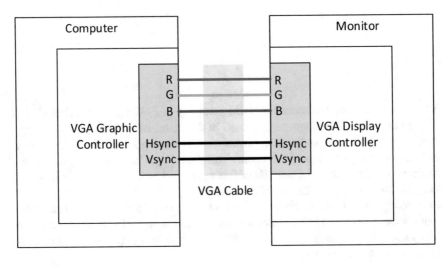

Hsync $\longrightarrow$ Horizontal Synchronization

Vsync $\longrightarrow$ Vertical Synchronization

Fig. 6.2 VGA connection lines

In Fig. 6.2, R, G, B lines carry analog data, on the other hand, "Hsync" and "Vsync" lines carry digital data. Inside VGA graphic controller, there is also "dena", display enable, and signal. The durations for the control signals "Hsync" and "Vsync" are controlled by the pixel clock generator, which has a value of 25.175 MHz for 640 × 480 pixel resolution. "Hsync" and "Vsync" control signals are used for the determination of the positions of new lines and new frames.

Internal part of the graphic controller is shown in Fig. 6.3. Inside the graphic controller, there are control signal generator, image generator, and digital and analog convert units. Timing of the color signals "R", "G", "B" is determined under the control of "Pixel Clk", "Hactive", "Vactive", and "Dena" signals.

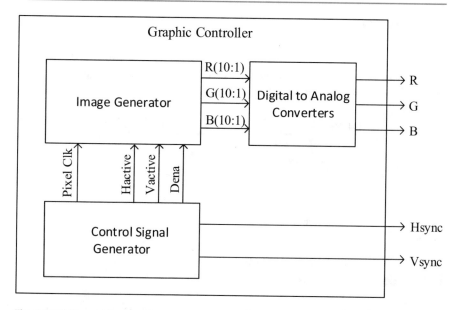

Fig. 6.3 VGA graphic controller

6.1.2   VGA Monitors

VGA monitors are cathode ray tubes (CRTs) as shown in Fig. 6.4. Using vertical and horizontal synchronization signals, the coordinates of the target of the cathode ray is determined. A VGA monitor takes "R", "G", "B" analog color information and digital horizontal and vertical synchronization signals from VGA graphic controller.

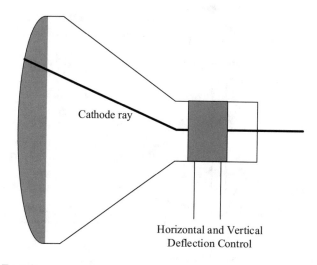

Fig. 6.4 CRT monitor

6.1.3 Pixel Clock

The pixels are displayed under the control of pixel clock. For 640×480 resolution with refresh rate 60 Hz, the pixel clock is 25.175 MHz for which the pixel period is almost 0.4 µs as depicted in Fig. 6.5.

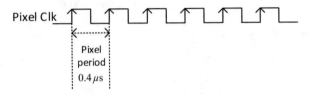

Fig. 6.5 Pixel clock pulse train

6.2   Basic VGA Format

In basic VGA format, the monitor screen is considered as a matrix of size 640×480 containing pixels in this matrix structure. That is, the monitor has 480 rows and there are 640 pixels in a row. A series of photos displayed on the monitor is nothing but videos watched on a monitor, the photos are displayed fast enough so that the human eye is unaware of the photo changes, and the human brain interprets the sequential displays as video display.

The screen is refreshed 60 times per second, i.e., the refresh rate of the screen is 60 Hz. The current flow graph on the screen is displayed in Fig. 6.6.

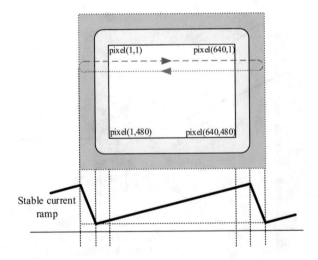

Fig. 6.6 The current flow graph on the screen

The current waveform passing through the monitor coils and the corresponding "Hsync" and "Hactive" signal waveforms are displayed in Fig. 6.7.

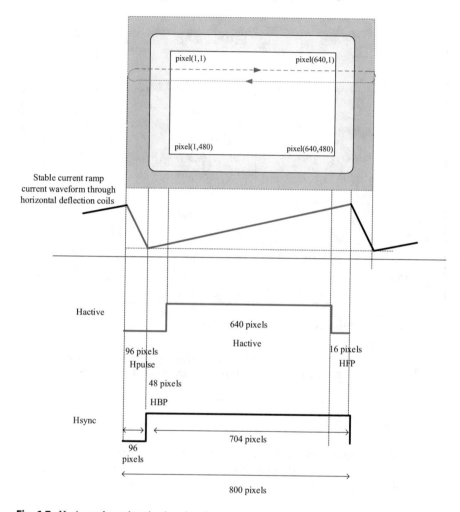

Fig. 6.7 Horizontal synchronization signal and current flow graphs

It is seen from Fig. 6.7 that the visible pixels are displayed on the screen for the "Hactive" portion of the "Hsync" signal. A better view showing the active region of the monitor screen and the corresponding control signals is depicted in Fig. 6.8.

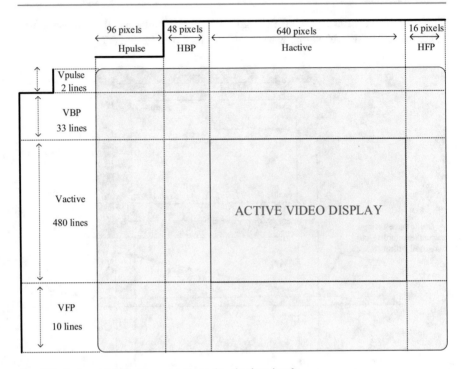

Fig. 6.8 Active video display area and synchronization signals

In Fig. 6.8, "Hsync" signal and its portions are displayed at the top of the screen, and "Vsync" signal and its portions are shown on the left side of the screen.

6.2.1 Hsync Signal

"Hsync" signal and its portions are depicted for 640 × 480 resolution in Fig. 6.9 in detail where it is seen that the pulse durations for "Hsync" signal are multiples of the period of the pixel clock.

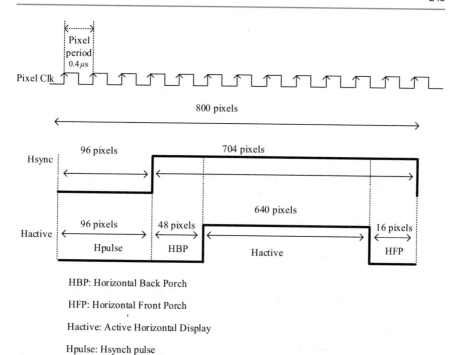

HBP: Horizontal Back Porch

HFP: Horizontal Front Porch

Hactive: Active Horizontal Display

Hpulse: Hsynch pulse

Fig. 6.9 Horizontal synchronization signal and its parts

6.2.2   Vsync Signal

"Vsync" signal and its portions are depicted for 640 × 480 resolution in Fig. 6.10 in detail.

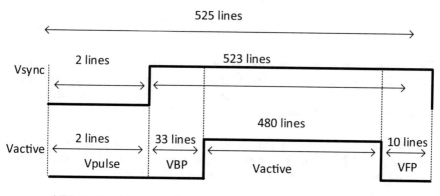

VBP: Vertical Back Porch

VFP: Vertical Front Porch

Vactive: Active Vertical Display

Vpulse: Vsynch pulse

Fig. 6.10 Vertical synchronization signal and its parts

In Fig. 6.11, the relationship between horizontal and vertical synchronization signals are explained.

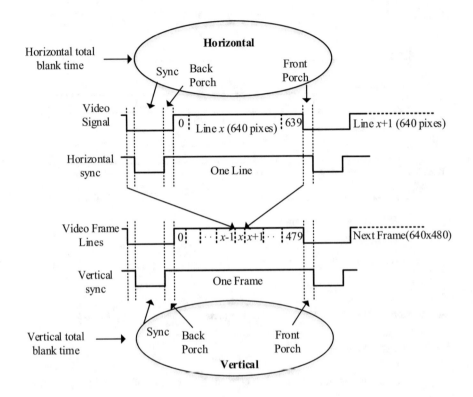

Fig. 6.11 Relationship between horizontal and vertical synchronization signals

Basic VGA protocol is defined for 640 × 480 pixels and contains five interface signals, i.e., horizontal synchronization, vertical synchronization, red signal, green signal, and blue signal. Among these signals, color signals are analogs while others are digital. Voltage levels in red-green-blue (RGB) determine what color the pixel has.

Images are formed by the periodic scan of the monitor screen by the electron beam. Figure 6.12 illustrates the scanning operation performed by the electron beam for the resolution of 640 × 480 display on a CRT monitor.

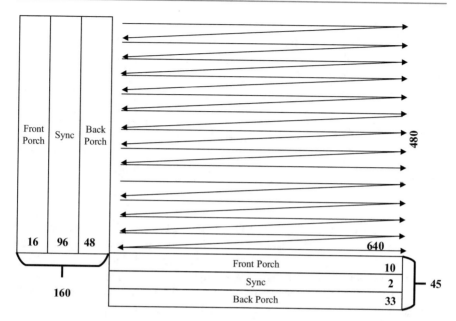

Fig. 6.12 Electron beam scan

6.2.3 VGA Resolution Modes

VGA standard supports different resolutions such as SVGA (800 × 600), WVGA (800 × 480), XVGA (1024 × 768), HD720 (1280 × 720), and HD1080 (1920 × 1080). For different resolutions, different scanning, waiting, and refreshing times are defined. In Table 6.1, information about different resolution modes is provided.

Table 6.1 VGA resolutions and timing parameters

Video timings	VGA 640 × 480 (60 Hz)	SVGA 800 × 600 (60 Hz)	HD1280 × 720-720p (60 Hz)
Pixel Clock	~25 MHz	40 MHz	74.25 MHz
TMDS Clock	~250 MHz	400 MHz	742.50 MHz
Horizontal Timings	*Duration in terms of pixels*	*Duration in terms of pixels*	*Duration in terms of pixels*
Active Pixels	640	800	1280
Front Porch	16	40	110
Snyc Width	96	128	40
Back Porch	48	88	220
Blanking Total	160	256	370
Total Pixels	800	1056	1650

(continued)

Table 6.1   (continued)

Video timings	VGA 640 × 480 (60 Hz)	SVGA 800 × 600 (60 Hz)	HD1280 × 720-720p (60 Hz)
Vertical Timings	*Duration in terms of pixels*	*Duration in terms of pixels*	*Duration in terms of pixels*
Active Lines	480	600	720
Front Porch	10	1	5
Snyc Width	2	4	5
Back Porch	33	23	20
Blanking Total	45	28	30
Total Lines	525	628	750

To display images or movies using VGA, horizontal and vertical synchronization signals are to be generated. Assume that original VGA (640 × 480) is going to be implemented. In VGA, 640 rows and 480 columns are needed to be scanned. Thus, horizontal synchronization signal is asserted after every time all columns are scanned, i.e., 640 rows. Besides, assertion of vertical synchronization signal is done when the beginning of scanning and at the end of completion of all rows.

6.3   VGA Connector

VGA connector shown in Fig. 6.13 is a 15-pin connector called DB15. The pins used in the display operation shown in Fig. 6.13 are red, green, and blue pins which carry analog signal varying from 0 to 0.7 V. The pins 4, 11, 12, and 15 are used for monitor identification and are not labeled in Fig. 6.13, since in this chapter we are only interested in image display operation.

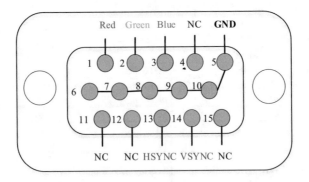

Fig. 6.13 VGA connector

6.4   VHDL Design for VGA Interface

An image on the monitor is formed by coloring the pixels. In Fig. 6.14, a single pixel is colored in red. The coordinates of a pixel are determined using "Hsync" and "Vsync" waveforms whose generations are controlled by pixel clock.

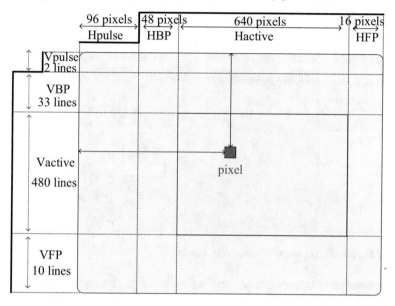

Fig. 6.14 Determination of pixel coordinates

Coloring a set of pixels, more complex pictures are formed as shown in Fig. 6.15.

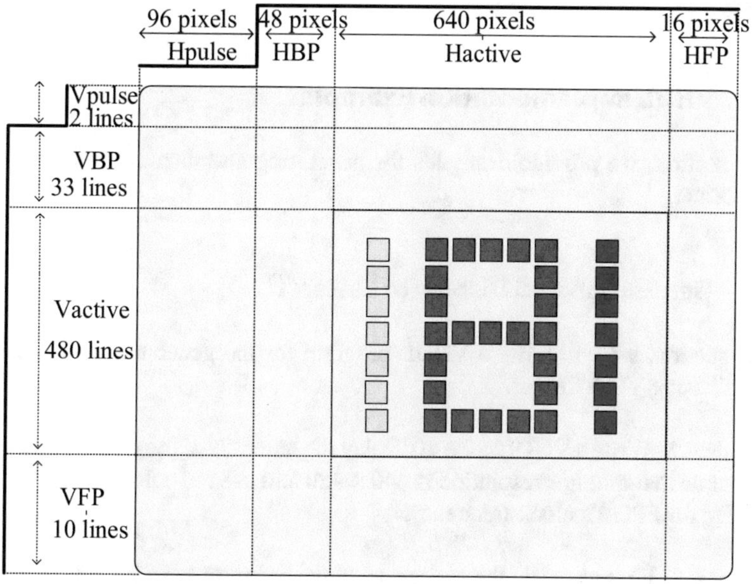

Fig. 6.15 Image formation using VGA format

Example 6.1 In Fig. 6.16, the coordinate calculations to display the letter "I" at the center of the monitor is illustrated.

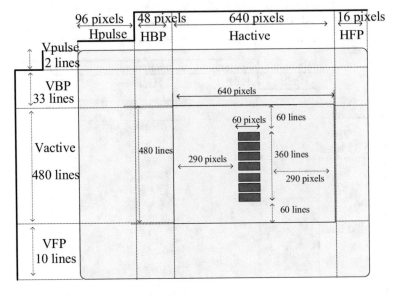

Fig. 6.16 Formation of letter "I" using VGA format

The generation of images can be done in several ways. The images to be displayed on the monitor screen can be generated by a VHDL code, or they can be read from memory units such as SRAM, EEPROM, or they can be retrieved from real-time image sources like video cameras. Now let us explain how to display images on a monitor screen using VHDL programming. We will explain the subject through examples.

6.5   VHDL Implementation Examples

In this section, we provide examples for generating and displaying shapes using VGA format.

6.5.1   Generation and Display of Letter "I"

In this section, we will write a VHDL program for the generation and display of letter "I" using VGA format.

Example 6.2 Write a VHDL code to display the letter "I" at the center of the monitor. Assume that monitor resolution is 640 × 480 and use red color for display. Take 100 MHz for FPGA's clock frequency.

Solution 6.2 First we write the entity part of the program as in PR 6.1.

```
library ieee;
use ieee.std_logic_1164.all;

entity vga_monitor is
   port(clk, reset: in std_logic;
        Hsync, Vsync: buffer std_logic;
        r, g, b: out std_logic;
        nblanck, nsync: out std_logic);
end vga_monitor;
```

PR 6.1   Program 6.1

We define constant integers objects to identify specific part of "Hsync" and "Vsync" waveforms in the declarative part of the architecture unit as in PR 6.2. Besides, signal objects are defined in PR 6.2 for clock generation and counting operations.

```
architecture Behavioral of vga_monitor is

   constant h1: integer:= 96; -- h_pulse
   constant h2: integer:= 144; -- h_pulse+hbp
   constant h3: integer:= 784; -- h_pulse + hbp +h_active
   constant h4: integer:= 800; -- h_pulse + hbp +h_active+hfp
   constant v1: integer:= 2; -- v_pulse
   constant v2: integer:= 35; -- v_pulse +vbp
   constant v3: integer:= 515; -- v_pulse +vbp+Vactive
   constant v4: integer:= 525; -- v_pulse +vbp+Vactive+vfp


   signal Hactive, Vactive, dena: std_logic;
   signal pixel_clk, pixel_clk1: std_logic;
   signal Vcount: positive range 1 to v4;
   signal Hcount: positive range 1 to h4;
begin
```

PR 6.2   Program 6.2

For resolution 640×480, the pixel clock frequency is 25 MHz which can be generated from 100 MHz FPGA's clock frequency using two simple processes in PR 6.3.

```
p1: process(clk) -- 100MHz ---> 50MHz
begin
  if(clk'event and clk='1') then
    pixel_clk1 <= not pixel_clk1;
  end if;
end process;

p2: process(pixel_clk1) --50MHz---> 25MHz
begin
  if(pixel_clk1'event and pixel_clk1='1') then
    pixel_clk<=not pixel_clk;
  end if;
end process;
```

PR 6.3   Program 6.3

Using pixel clock source, we can generate the "Hsync" signal using the process in PR 6.4.

```
--Hsync signal generation:
p3: process(pixel_clk)
begin
  if(pixel_clk'event and pixel_clk='1') then
    Hcount<=Hcount + 1;
    if(Hcount=h1) then
      Hsync<='1';
    elsif(Hcount=h2) then
      Hactive<='1';
    elsif(Hcount=h3) then
      Hactive<= '0';
    elsif(Hcount=h4) then
      Hsync<='0';
      Hcount<=1;
    end if;
  end if;
end process;
```

PR 6.4   Program 6.4

Using "Hsync" signal, we can generate the "Vsync" signal using the process in PR 6.5.

```
--Vsync signal generation:
p4: process(Hsync)
begin
  if(Hsync'event and Hsync='0') then
    Vcount<=Vcount + 1;
    if(Vcount=v1) then
      Vsync<='1';
    elsif(Vcount=v2) then
      Vactive<='1';
    elsif(Vcount=v3) then
      Vactive<='0';
    elsif(Vcount=v4) then
      Vsync<='0';
      Vcount<=1;
    end if;
  end if;
end process;
```

PR 6.5   Program 6.5

Using the coordinate information shown in Fig. 6.16, we can generate the required image as in PR 6.6.

```
-- Image generator
Img_gen: process(pixel_clk)
begin
  if(dena='1') then
    if(Hcount>=h2+290 and Hcount<=h2+290+60 and Vcount>=v2+60 and Vcount<=v2+60+360)
then
      r<='1'; g<='0'; b<='0';
    else
      r<='0'; g<='0'; b<='0';
    end if;
  end if;
```

PR 6.6 Program 6.6

Combining all the program units, we get the overall program as in PR 6.7.

```
library ieee;
use ieee.std_logic_1164.all;

entity vga_monitor is
  port(clk, reset: in std_logic;
       Hsync, Vsync: buffer std_logic;
       r, g, b: out std_logic;
       nblanck, nsync: out std_logic);
end vga_monitor;

architecture logic_flow of vga_monitor is

  constant h1: integer:= 96;
  constant h2: integer:= 144;
  constant h3: integer:= 784;
  constant h4: integer:= 800;
  constant v1: integer:= 2;
  constant v2: integer:= 35;
  constant v3: integer:= 515;
  constant v4: integer:= 525;

  signal Hactive, Vactive, dena: std_logic;
  signal pixel_clk, pixel_clk1: std_logic;
  signal Vcount: positive range 1 to v4;
  signal Hcount: positive range 1 to h4;
begin
--- Display enable generation:
dena<= Hactive and Vactive;

-- Static signals for DACs:
nblanck<='1'; --no direct blanking
nsync<='0'; --no sync on green

p1: process(clk)--100MHz ---> 50MHz
begin
  if(clk'event and clk='1') then
    pixel_clk1<=not pixel_clk1;
  end if;
end process;

p2: process(pixel_clk1)-- ---> 25MHz
begin
  if(pixel_clk1'event and pixel_clk1='1') then
    pixel_clk<=not pixel_clk;
  end if;
end process;
```

```
--Hsync signal generation:
p3: process(pixel_clk)
begin
  if(pixel_clk'event and pixel_clk='1') then
    Hcount<=Hcount + 1;
    if(Hcount=h1) then
      Hsync<='1';
    elsif(Hcount=h2) then
      Hactive<='1';
    elsif(Hcount=h3) then
      Hactive<='0';
    elsif(Hcount=h4) then
      Hsync<='0';
      Hcount<=1;
    end if;
  end if;
end process;

--Vsync signal generation:
p4: process(Hsync)
begin
  if(Hsync'event and Hsync='0') then
    Vcount<=Vcount + 1;
    if(Vcount=v1) then
      Vsync<='1';
    elsif(Vcount=v2) then
      Vactive<='1';
    elsif(Vcount=v3) then
      Vactive<='0';
    elsif(Vcount=v4) then
      Vsync<='0';
      Vcount<=1;
    end if;
  end if;
end process;
-- Image generator
image_gen: process(pixel_clk)
begin
  if(dena='1') then
    if(Hcount>=h2+290 and Hcount<=h2+290+60
      and Vcount>=v2+60 and Vcount<=v2+60+360)
then
      r<='1'; g<='0'; b<='0';
    else
      r<='0'; g<='0'; b<='0';
    end if;
  end if;
end process;  end logic_flow;
```

PR 6.7 Program 6.7

6.5.2   Generation and Display of Square Shape

In this section, we will write a VHDL program for the generation and display of a square shape using VGA format.

Example 6.3 Write a VHDL program to display a square of size 100×100 pixel on the center of the monitor screen. The color of the square is red. Resolution of the monitor connected to FPGA board via a VGA cable is 640×480. The colors of pixels are determined using 12 bits in which 4 bits are used for red color, the other 4 bits are used for green, and lastly the last 4 bits are used for blue. The clock frequency of the FPGA device is 100 MHz.

Solution 6.3 We will write the VHDL program using components. The components to be used in VHDL program are depicted in Fig. 6.17.

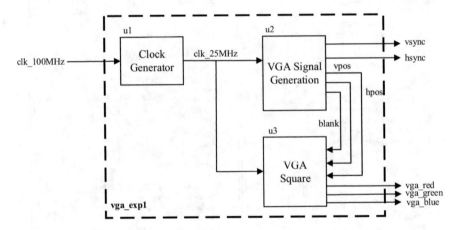

Fig. 6.17 Components of VHDL program

As it is seen from Fig. 6.17, we have three components to be implemented. One is used for the generation of pixel clock which is 25 MHz for 640×480 resolution. The second component is used to generate the control signals such as "vsync", "hsync", and counter parameters "vpos", "hpos", and blank control signal. The image generation is performed by the third component called "VGA square" in Fig. 6.17.

The entity part of the VHDL program is written in PR 6.8 where control and color signals are defined.

```
library ieee;
use ieee.std_logic_1164.all;
use ieee.numeric_std.all;

entity vga_exp1 is
  port(clk_100MHz: in std_logic;
       hsync,vsync: out std_logic;
       vgaRed,vgaGreen,vgaBlue : out std_logic_vector(3 downto 0));
end vga_exp1;
```

PR 6.8 Program 6.8

Signal objects for 25 MHz pixel clock and counter parameters are defined in the declarative part of the architecture in PR 6.9.

```
architecture logic_flow of vga_exp1 is

signal clk_25MHz: std_logic;
signal blank: std_logic:= '0';
signal hpos, vpos: positive range 1 to 1024;
```

PR 6.9 Program 6.9

We need to add three components to the declarative part of the architecture. The component for pixel clock generator can be written as in PR 6.10.

```
component clock_generator is
  port (clk_100MHz: in std_logic;
        clk_25MHz: out std_logic);
end component;
```

PR 6.10 Program 6.10

The component for the generation of the control signals can be written as in PR 6.11.

```
component vga_signal_gen is
  port (clk: in  std_logic;
        blank: out std_logic;
        hsync,vsync: out std_logic;
        hpos,vpos: out positive range 1 to 1024);
end component;
```

PR 6.11   Program 6.11

And finally, the component for the generation of the required image is written as in PR 6.12.

```
component vga_square is
port (clk: in std_logic;
      blank_in: in std_logic;
      hpos, vpos: in positive range 1 to 1024;
      vga_red, vga_green, vga_blue: out std_logic_vector(3 downto 0);
      vga_blank: out std_logic);
end component;
```

PR 6.12   Program 6.12

The components are instantiated in the body of the architecture using the port map function as in PR 6.13.

```
begin
  u1: clock_generator port map(              u3: vga_square port map(
      clk_100MHz => clk_100MHz,                  clk        => clk_25MHz,
      clk_25MHz  => clk_25MHz);                  blank_in  => blank,
                                                 hpos       => hpos,
  u2: vga_signal_gen port map(                   vpos       => vpos,
      clk        => clk_25MHz,                    vga_red   => vgaRed,
      blank      => blank,                        vga_green => vgaGreen,
      hsync      => hsync,                        vga_blue  => vgaBlue);
      vsync      => vsync,
      hpos       => hpos,                     end logic_flow;
      vpos       => vpos);
```

PR 6.13   Program 6.13

Combining all the program parts, we obtain the main VHDL program as in PR 6.14.

```
library ieee;                                              vga_blank: out std_logic);
use ieee.std_logic_1164.all;                             end component;
use ieee.numeric_std.all;
                                                         begin
entity vga_exp1 is                                       u1: clock_generator port map(
  port(clk_100MHz: in std_logic;                             clk_100MHz => clk_100MHz,
       hsync, vsync: out std_logic;                          clk_25MHz  => clk_25MHz);
       vgaRed, vgaGreen, vgaBlue: out
               std_logic_vector(3 downto 0));            u2: vga_signal_gen port map(
end vga_exp1;                                                 clk      => clk_25MHz,
                                                             blank    => blank,
                                                             hsync    => hsync,
architecture logic_flow of vga_exp1 is                       vsync    => vsync,
                                                             hpos     => hpos,
  signal clk_25MHz: std_logic;                               vpos     => vpos);
  signal blank: std_logic:= '0';
  signal hpos, vpos: positive range 1 to 1024;          u3: vga_square port map(
  component clock_generator is                               clk      => clk_25MHz,
    port(clk_100MHz: in std_logic;                           blank_in => blank,
         clk_25MHz: out std_logic);                          hpos     => hpos,
  end component;                                             vpos     => vpos,
  component vga_signal_gen is                               vga_red  => vgaRed,
    port(clk: in  std_logic;                                vga_green => vgaGreen,
         blank: out std_logic;                              vga_blue => vgaBlue);
         hsync, vsync: out std_logic;
         hpos,vpos: out positive range 1 to 1024);      end logic_flow;
  end component;
  component vga_square is
    port(clk: in std_logic;
         blank_in: in std_logic;
         hpos, vpos: in positive range 1 to 1024;
         vga_red, vga_green, vga_blue: out
                 std_logic_vector(3 downto 0);
```

PR 6.14   Program 6.14

After writing the main program as in PR 6.14, we can start writing the VHDL programs for the components used in the main program. The VHDL program for the component "clock_generator" can be written as in PR 6.15.

```vhdl
library ieee;
use ieee.std_logic_1164.all;
use ieee.numeric_std.all;

entity clock_generator is
  port(clk_100MHz: in std_logic;
       clk_25MHz: out std_logic);
end clock_generator;

architecture logic_flow of clock_generator is
  signal count: integer range 0 to 3:=0;
  signal signal_25MHz: std_logic:='0';
begin
  clk_25MHz<=signal_25MHz;

  clk25MHz: process(clk_100MHz)
  begin
    if(rising_edge(clk_100MHz)) then
      if(count=3) then
        signal_25MHz<=not signal_25MHz;
        count<=0;
      else
        count<=count + 1;
      end if;
    end if;
  end process;
end logic_flow;
```

PR 6.15   Program 6.15

Next, we explain how to write the VHDL program for the component "vga_sig-nal_gen" which is used to generate the control signals, such as horizontal synchronization and vertical synchronization.

In PR 6.16, the entity part is written and in the declarative part of the architecture signal objects "x" and "y", to be used for counter indices during the generation of synchronization signals, are defined. Similarly, signal objects "act_pxl_hrzntl" and "act_pxl_vrtc" are defined for counter indices for the generation of active part of the synchronization signals.

```
library ieee;
use ieee.std_logic_1164.all;
use ieee.numeric_std.all;

entity vga_signal_gen is
  port(clk: in std_logic;
       blank: out std_logic;
       hsync, vsync: out std_logic;
       hpos, vpos: out positive range 1 to 1024);
end vga_signal_gen;

architecture logic_flow of vga_signal_gen is
  signal x, y: integer range 0 to 1023:=0;
  signal act_pxl_hrzntl, act_pxl_vrtcl: positive range 1 to 1024:=1;
  signal hsync_sig: std_logic:='0';
  signal Hactive,Vactive: std_logic:='0';
begin
```

PR 6.16 Program 6.16

The generations of the horizontal synchronization and horizontal active signals are implemented in PR 6.17, PR 6.18, PR 6.19, PR 6.20 and explained in a stepwise manner for monitor resolution 640 × 480.

```
horizontal_sync: process(clk)
begin
  if(rising_edge(clk)) then
    x<=x + 1;
    if (x<96) then
      hsync_sig<='0'; Hactive<='0';
```

PR 6.17 Program 6.17

```
horizontal_sync: process(clk)
begin
  if(rising_edge(clk)) then
    x <=x + 1;
    if(x<96) then
      hsync_sig<='0'; Hactive<='0';
    elsif (x>=96 and x<144) then
      hsync_sig<='1'; Hactive<='0';
```

PR 6.18 Program 6.18

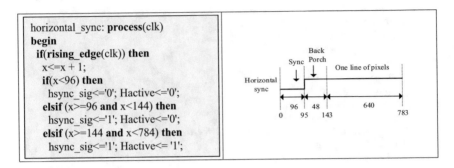

```
horizontal_sync: process(clk)
begin
 if(rising_edge(clk)) then
  x<=x + 1;
  if(x<96) then
   hsync_sig<='0'; Hactive<='0';
  elsif (x>=96 and x<144) then
   hsync_sig<='1'; Hactive<='0';
  elsif (x>=144 and x<784) then
   hsync_sig<='1'; Hactive<= '1';
```

PR 6.19   Program 6.19

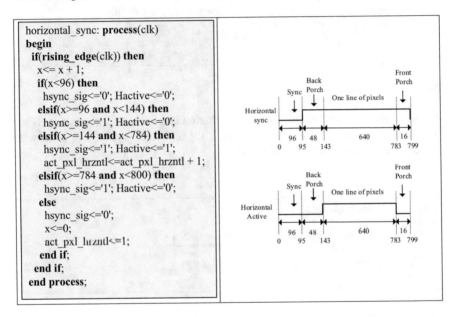

```
horizontal_sync: process(clk)
begin
 if(rising_edge(clk)) then
  x<= x + 1;
  if(x<96) then
   hsync_sig<='0'; Hactive<='0';
  elsif(x>=96 and x<144) then
   hsync_sig<='1'; Hactive<='0';
  elsif(x>=144 and x<784) then
   hsync_sig<='1'; Hactive<='1';
   act_pxl_hrzntl<=act_pxl_hrzntl + 1;
  elsif(x>=784 and x<800) then
   hsync_sig<='1'; Hactive<='0';
  else
   hsync_sig<='0';
   x<=0;
   act_pxl_hrzntl<=1;
  end if;
 end if;
end process;
```

PR 6.20   Program 6.20

Generation of vertical synchronization signal can be done in a similar manner. It is important to state that vertical synchronization signals are generated at every rising edge of the horizontal synchronization signal. The generation of vertical synchronization signal is done and explained in a stepwise manner in program parts from PR 6.21, PR 6.22, PR 6.23 and PR 6.24.

```
vertical_sync: process(hsync_sig)
begin
 if(rising_edge(hsync_sig)) then
  y<=y + 1;
  if(y<2) then
   Vsync<= '0'; Vactive<='0';
```

PR 6.21   Program 6.21

```
vertical_sync: process(hsync_sig)
begin
  if(rising_edge(hsync_sig)) then
    y<=y + 1;
    if (y<2) then
      Vsync<='0';Vactive<='0';
    elsif (y>=2 and y<35) then
      Vsync<='1';Vactive<='0';
```

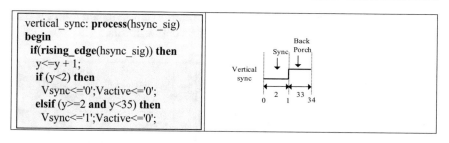

PR 6.22   Program 6.22

```
vertical_sync: process(hsync_sig)
begin
  if(rising_edge(hsync_sig)) then
    y<=y + 1;
    if(y<2) then
      Vsync<='0'; Vactive<='0';
    elsif(y>=2 and y<35) then
      Vsync<='1'; Vactive<='0';
    elsif(y>=35 and y<515) then
      Vsync<='1';Vactive<='1';
```

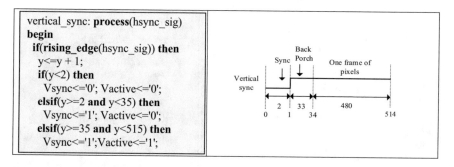

PR 6.23   Program 6.23

```
vertical_sync: process (hsync_sig)
begin
  if(rising_edge(hsync_sig)) then
    y<=y + 1;
    if (y<2) then
      Vsync<='0'; Vactive<='0';
    elsif (y>=2 and y<35) then
      Vsync<='1'; Vactive<='0';
    elsif (y>=35 and y<515) then
      Vsync<='1'; Vactive<='1';
      act_pxl_vrtcl<= act_pxl_vrtcl + 1;
    elsif (y>=515 and y<525) then
      Vsync<='1'; Vactive<='0';
    else
      Vsync<='0';
      y<=0;
      act_pxl_vrtcl<=1;
    end if;
  end if;
end process; end logic_flow;
```

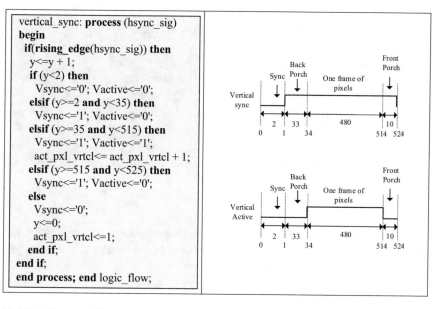

PR 6.24   Program 6.24

Combining all the program parts, we get the overall VHDL program as in PR 6.25 for the component used for the generation of the control signals.

```vhdl
library ieee;
use ieee.std_logic_1164.all;
use ieee.numeric_std.all;

entity vga_signal_gen is
  port(clk: in std_logic;
       blank: out std_logic;
       hsync,vsync: out std_logic;
       hpos,vpos: out positive range 1 to 1024);
end vga_signal_gen;

architecture logic_flow of vga_signal_gen is
  signal x, y: integer range 0 to 1023:=0;
  signal act_pxl_hrzntl,act_pxl_vrtcl: positive
                               range 1 to 1024:=1;
  signal hsync_sig: std_logic:='0';
  signal Hactive, Vactive: std_logic:='0';

begin
  hpos<=act_pxl_hrzntl;
  vpos<=act_pxl_vrtcl;
  Hsync<=hsync_sig;
  blank<=not(Hactive and Vactive);
  --Generation of Horizontal signals
  horizontal_sync: process (clk)
  begin
    if(rising_edge(clk)) then
      x<=x + 1;
      if(x<96) then
        hsync_sig<='0'; Hactive<='0';
      elsif (x>=96 and x<144) then
        hsync_sig<='1'; Hactive<='0';
      elsif (x>=144 and x<784) then
        hsync_sig<='1'; Hactive<='1';
        act_pxl_hrzntl<=act_pxl_hrzntl + 1;
      elsif(x>=784 and x<800) then
        hsync_sig<='1'; Hactive<='0';
      else
        hsync_sig<='0';
        x<=0;
        act_pxl_hrzntl<=1;
      end if;
    end if;
  end process;

  --Generation of Vertical signals
  vertical_sync: process (hsync_sig)
  begin
    if(rising_edge(hsync_sig)) then
      y<=y + 1;
      if(y<4) then
        Vsync<='0'; Vactive<='0';
      elsif (y>=4 and y<27) then
        Vsync<='1'; Vactive<='0';
      elsif (y>=27 and y<627) then
        Vsync<='1'; Vactive<='1';
        act_pxl_vrtcl<=act_pxl_vrtcl + 1;
      elsif (y>=627 and y<628) then
        Vsync<='1'; Vactive<='0';
      else
        Vsync<='0';
        y<=0;
        act_pxl_vrtcl<=1;
      end if;
    end if;
  end process;
end logic_flow;
```

PR 6.25   Program 6.25

Now we will write the VHDL program for the image generation component "vga_square" which is used to form a square on the center of the monitor. The entity part of the VHDL program is written in PR 6.26.

```
library ieee;
use ieee.std_logic_1164.all;
use ieee.numeric_std.all;

entity vga_square is
  port(clk: in std_logic;
       blank_in: in std_logic;
       hpos, vpos: in positive range 1 to 1024;
       vga_red, vga_green, vga_blue: out std_logic_vector(3 downto 0));
end vga_square;
```

PR 6.26   Program 6.26

In PR 6.27, counter indices, and size parameter for the square shape to be drawn are defined in the declarative part of the architecture unit. Size parameter for the square shape is initialized to half of the real size. This is due to square drawing logic to be employed in the VHDL code.

```
architecture logic_flow of vga_square is

  signal size: positive range 1 to 1024:=100;
  signal obj_X_pos: positive range 1 to 1024:=320;
  signal obj_Y_pos: positive range 1 to 1024:=240;

begin
```

PR 6.27   Program 6.27

In the body of the architecture unit, we need a process to draw the square shape. The template for this process is depicted in PR 6.28.

```
architecture logic_flow of vga_square is
begin
  square_draw: process(clk)
  begin
    if(rising_edge(clk)) then
      if(blank_in='0') then
        --
        --Draw your shape here
        --
      end if;
    end if;
  end process;
end logic_flow;
```

PR 6.28   Program 6.28

Considering the coordinates of the corners of the square shape, the coordinates of the pixels in the square region can be determined using an **if** statement with four different conditions as in PR 6.29. The four regions shown in PR 6.29 can be indicated by four different Boolean conditions of PR 6.29. Intersection of the Boolean conditions define the square region.

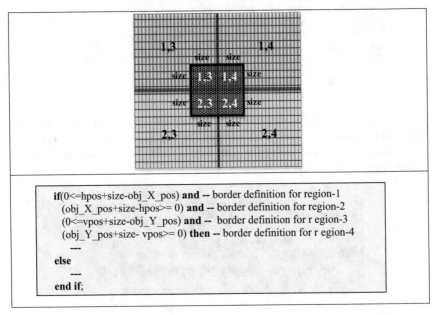

```
if(0<=hpos+size-obj_X_pos) and -- border definition for region-1
  (obj_X_pos+size-hpos>= 0) and -- border definition for region-2
  (0<=vpos+size-obj_Y_pos) and -- border definition for region-3
  (obj_Y_pos+size- vpos>= 0) then -- border definition for region-4

  ---
else
  ---
end if;
```

PR 6.29 Program 6.29

Using the coordinates of pixels inside the square shape, we can color square shape as in PR 6.30.

```
square_draw: process(clk)
begin
if(rising_edge(clk)) then
  if(blank_in='0') then
    if((0<=hpos + size - obj_X_pos) and
      (obj_X_pos + size-hpos>= 0) and
      (0<=vpos + size-obj_Y_pos) and
      (obj_Y_pos + size- vpos>= 0)) then
      vga_red<=x"f";
      vga_green<=x"0";
      vga_blue<=x"0";
    else
      vga_red<=x"f";
      vga_green<=x"f";
      vga_blue<=x"f";
    end if;
  end if;
end if;
end process;
```

PR 6.30 Program 6.30

Combining all program units, we obtain the VHDL program for the component "vga_square" as in PR 6.31.

```vhdl
library ieee;
use ieee.std_logic_1164.all;
use ieee.numeric_std.all;

entity vga_square is
  port(clk: in std_logic;
       blank_in: in std_logic;
       hpos, vpos: in positive range 1 to 1024;
       vga_red, vga_green, vga_blue: out
                 std_logic_vector(3 downto 0));
end vga_square;

architecture logic_flow of vga_square is

  signal size: positive range 1 to 1024:=100;
  signal obj_X_pos: positive range 1 to 1024:=320;
  signal obj_Y_pos: positive range 1 to 1024:=240;
begin
  square_draw :process(clk)
  begin

    if(rising_edge(clk)) then
      if(blank_in='0') then
        if((0<= hpos + size - obj_X_pos) and
           (obj_X_pos + size-hpos>= 0) and
           (0<= vpos + size-obj_Y_pos) and
           (obj_Y_pos + size- vpos>= 0)) then
          vga_red<= x"f";
          vga_green<=x"0";
          vga_blue<=x"0";
        else
          vga_red<=x"f";
          vga_green<=x"f"; vga_blue<=x"f";
        end if;
      else
        vga_red<=x"0"; vga_green<=x"0";
        vga_blue<=x"0";
      end if;
    end if;
  end process;
end logic_flow;
```

PR 6.31 Program 6.31

In Fig. 6.18, a typical square shape drawn for VGA resolution 640 × 480 by the VHDL program in PR 6.31 is depicted.

Fig. 6.18 A square at the center of the screen

6.5.3 Generation and Display of Moving Square

In this section, we will explain how to develop a VHDL program to move a geometric shape on the monitor screen. We will explain the subject with an example.

Example 6.4 Write a VHDL program which displays a square with sides of 8 pixels long at the center of the monitor. This square shape can move in four directions, up, down, left, and right, controlled by four buttons. Pressing a button generates logic "1". Resolution of the monitor connected to FPGA board via a VGA cable is 640 × 480. Use 12 bits to color a pixel, and 4 of these bits are used for red color, the other 4 are used for green, and the last 4 bits are used for blue.

Solution 6.4 This example can be considered an improved version of the previous example. Our design consists of three components as shown in Fig. 6.19. The only difference from the previous example appears in the design of third component, i.e., u3. The component u3 has 4four input bits for the direction of motion, and it has three output bits.

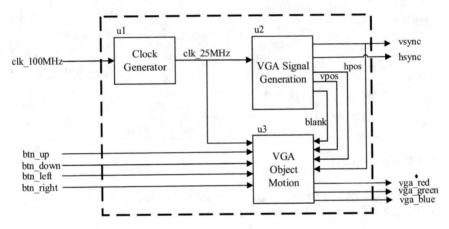

Fig. 6.19 Components of VHDL program for Example 6.4

The entity part of the VHDL program can be written as in PR 6.32 where different from the previous example input ports for directions of motions are defined.

```
library ieee;
use ieee.std_logic_1164.all;
use ieee.numeric_std.all;

entity vga_exp2 is
  port(clk_100MHz: in std_logic;
       hsync, vsync: out std_logic;
       btn_left, btn_right: in std_logic;
       btn_up, btn_down: in std_logic;
       vgaRed, vgaGreen, vgaBlue: out std_logic_vector(3 downto 0));
end vga_exp2;
```

PR 6.32 Program 6.32

The signal objects defined in the declarative part of the architecture unit are the same as in the previous example, and they are given in PR 6.33.

```
architecture logic_flow of vga_exp2 is

signal clk_25MHz: std_logic;
signal blank, vsync_signal: std_logic:= '0';
signal hpos, vpos: positive range 1 to 1024;
    ---
    ---
```

PR 6.33 Program 6.33

Two components "u1" and "u2", written for the generations of pixel clock and control signals, used in the declarative part of the architecture unit of the main program are the same as in the previous example, and they are given in PR 6.34.

```
component clock_generator is
  port(clk_100MHz: in std_logic;
       clk_25MHz: out std_logic);
end component;

component vga_signal_gen is
  port(clk: in  std_logic;
       blank: out std_logic;
       hsync, vsync: out std_logic;
       hpos, vpos: out positive range 1 to 1024);
end component;
```

PR 6.34 Program 6.34

The third component, different from the previous example, to be used in the declarative part of the architecture unit of the main program, is given in PR 6.35. The third component is used for image generation and object moving.

```
component vga_obj_motion is
  port(clk: in std_logic;
       blank_in, vsync_in: in std_logic;
       btn_left, btn_right: in std_logic;
       btn_up, btn_down: in std_logic;
       hpos, vpos: in positive range 1 to 1024;
       vga_red, vga_green, vga_blue: out std_logic_vector(3 downto 0);
       vga_blank: out std_logic);
end component;
```

PR 6.35 Program 6.35

Using all the component units, we can write the main program as in PR 6.36.

```vhdl
library ieee;
use ieee.std_logic_1164.all;
use ieee.numeric_std.all;

entity vga_exp2 is
  port(clk_100MHz: in std_logic;
       hsync,vsync: out std_logic;
       btn_left, btn_right: in std_logic;
       btn_up, btn_down: in std_logic;
       vgaRed,vgaGreen,vgaBlue: out
              std_logic_vector(3 downto 0));
end vga_exp2;

architecture logic_flow of vga_exp2 is

  signal clk_25MHz: std_logic;
  signal blank, vsync_signal: std_logic:= '0';
  signal hpos, vpos: positive range 1 to 1024;

  component clock_generator is
    port(clk_100MHz: in std_logic;
         clk_25MHz: out std_logic);
  end component;

  component vga_signal_gen is
    port(clk: in  std_logic;
         blank: out std_logic;
         hsync, vsync: out std_logic;
         hpos, vpos: out positive range 1 to 1024);
  end component;

  component vga_obj_motion is
    port(clk: in std_logic;
         blank_in, vsync_in: in std_logic;
         btn_left, btn_right: in std_logic;
         btn_up, btn_down: in std_logic;
         hpos, vpos: in positive range 1 to 1024;
         vga_red, vga_green, vga_blue: out
                std_logic_vector(3 downto 0));

         vga_blank: out std_logic);
  end component;

begin
  u1: clock_generator port map(
         clk_100MHz => clk_100MHz,
         clk_25MHz  => clk_25MHz);

  u2: vga_signal_gen port map(
         clk        => clk_25MHz,
         blank      => blank,
         hsync      => hsync,
         vsync      => vsync_signal,
         hpos       => hpos,
         vpos       => vpos);

  u3: vga_obj_motion port map(
         clk        => clk_25MHz,
         blank_in   => blank,
         vsync_in   => vsync_signal,
         hpos       => hpos,
         vpos       => vpos,
         btn_left   => btn_left,
         btn_right  => btn_right,
         btn_up     => btn_up,
         btn_down   => btn_down,
         vga_red    => vgaRed,
         vga_green  => vgaGreen,
         vga_blue   => vgaBlue);

         vsync<=vsync_signal;
end logic_flow;
```

PR 6.36 Program 6.36

After writing the main program as in PR 6.36, we can start writing the VHDL programs for the components used in the main program.

Now, we will write the VHDL program for the third component "vga_obj_motion". The entity part and signal definitions in the declarative part of the architecture are given in PR 6.37. Signal objects are to be used for the coordinates of the square shape and for displacement.

```
library ieee;
use ieee.std_logic_1164.all;
use ieee.numeric_std.all;

entity vga_obj_motion is
  port(clk: in std_logic;
       blank_in, vsync_in: in std_logic;
       hpos, vpos: in positive range 1 to 1024;
       btn_left, btn_up, btn_right, btn_down: in std_logic;
       vga_red, vga_green, vga_blue: out std_logic_vector(3 downto 0));
end vga_obj_motion;

architecture logic_flow of vga_obj_motion is

  signal Size: positive range 1 to 1024:=8;
  signal obj_X_pos: positive range 1 to 1024:=320;
  signal obj_Y_pos: positive range 1 to 1024:=240;
  signal obj_X_motion: integer range -8 to 8:=0;
  signal obj_Y_motion: integer range -8 to 8:=0;
```

PR 6.37   Program 6.37

We need two processes in the architecture body. One of the processes, named as "obj_create", is used to draw the square shape and it is given in PR 6.38.

```
begin
  obj_create:process(clk)
  begin
    if(rising_edge(clk)) then
      if(blank_in='0') then
        if((0<= hpos + Size - obj_X_pos) and
        (obj_X_pos + Size-hpos>= 0) and
        (0<= vpos + Size-obj_Y_pos) and
        (obj_Y_pos + Size- vpos>= 0)) then
        vga_red<=x"f";
        vga_green<=x"0";
        vga_blue<=x"0";
      else

        vga_red<=x"f";
        vga_green<=x"f";
        vga_blue<=x"f";
        end if;
      else
        vga_red<=x"0";
        vga_green<=x"0";
        vga_blue<=x"0";
      end if;
    end if;
  end process;
```

PR 6.38   Program 6.38

The second process, named as "obj_move", is used for the displacement of the square shape. The implementation of the second process is given in PR 6.39.

```
obj_move: process  (vsync_in)
begin
if(rising_edge(vsync_in)) then
  -------- y axis motion --------
  if (btn_down='1' and btn_up='1')  then
    obj_Y_motion<=0;
  elsif (btn_up='0' and btn_down='1') then
    obj_Y_motion<=8;
  elsif (btn_down='0' and btn_up='1') then
    obj_Y_motion<=-8;
  elsif (btn_down='0' and btn_up='0') then
    obj_Y_motion<=0;
  end if;
  -------- x axis motion --------
  if (btn_left='1' and btn_right='1')  then
    obj_X_motion<=0;
  elsif (btn_left='0' and btn_right='1') then
    obj_X_motion<=-8;
  elsif (btn_right='0' and btn_left='1') then
    obj_X_motion<=8;
  elsif (btn_left='0' and btn_right='0') then
    obj_X_motion<=0;
  end if;
    obj_Y_pos<=obj_Y_pos + obj_Y_motion;
    obj_X_pos<=obj_X_pos + obj_X_motion;
  end process;
end logic_flow;
```

PR 6.39 Program 6.39

Combining all program parts, we get the overall code as in PR 6.40.

```vhdl
library ieee;
use ieee.std_logic_1164.all;
use ieee.numeric_std.all;

entity vga_obj_motion is
  port(clk: in std_logic;
       blank_in, vsync_in: in std_logic;
       hpos, vpos: in positive range 1 to 1024;
       btn_left, btn_right: in std_logic;
       btn_up, btn_down: in std_logic;
       vga_red, vga_green, vga_blue: out
                 std_logic_vector(3 downto 0));
end vga_obj_motion;

architecture logic_flow of vga_obj_motion is

  signal size: positive range 1 to 1024:=8;
  signal obj_X_pos: positive range 1 to 1024:=320;
  signal obj_Y_pos: positive range 1 to 1024:=240;
  signal obj_X_motion: integer range -8 to 8:=0;
  signal obj_Y_motion: integer range -8 to 8:=0;
begin
  obj_create: process(clk)
  begin
   if(rising_edge(clk)) then
    if(blank_in='0') then
     if((0<= hpos + size - obj_X_pos) and
        (obj_X_pos + size-hpos>=0) and
        (0<=vpos + size-obj_Y_pos) and
        (obj_Y_pos + size- vpos>=0)) then
     vga_red<=x"f";
     vga_green<=x"0";
     vga_blue<=x"0";
    else
     vga_red<=x"f";
     vga_green<=x"f";
     vga_blue<=x"f";
    end if;
```

```vhdl
    else
     vga_red<=x"0";
     vga_green<=x"0";
     vga_blue<=x"0";
    end if;
   end if;
  end process;
  obj_move: process (vsync_in)
  begin
  if(rising_edge(vsync_in)) then
-------- y axis motion --------
    if(btn_down='1' and btn_up='1') then
     obj_Y_motion<=0;
    elsif(btn_up='0' and btn_down='1') then
     obj_Y_motion<=8;
    elsif(btn_down='0' and btn_up='1') then
     obj_Y_motion<=-8;
    elsif(btn_down='0' and btn_up='0') then
     obj_Y_motion<=0;
    end if;
-------- x axis motion --------
    if(btn_left='1' and btn_right='1') then
     obj_X_motion<=0;
    elsif(btn_left='0' and btn_right='1') then
     obj_X_motion<=-8;
    elsif(btn_right='0' and btn_left='1')then
     obj_X_motion<=8;
    elsif(btn_left='0' and btn_right='0') then
     obj_X_motion<=0;
    end if;
    obj_Y_pos<=obj_Y_pos + obj_Y_motion;
    obj_X_pos<=obj_X_pos + obj_X_motion;
   end if;
  end process;
end logic_flow;
```

PR 6.40   Program 6.40

6.5.4   Generation and Display of a Filled-Circle and a Ring

In this section, we will show how to draw circles and rings using VHDL. We will explain the subject through an example.

Example 6.5 Write a VHDL program to draw one filled-circle and one ring shape with different radii as shown in Fig. 6.20. Use blue color for the circle and ring. The length of the radius of the filled-circle and inner radius of the ring is 100 pixels in length, and the length of the outer radius of the ring is 130 pixels in length. The resolution of the display is 800 × 600. The clock frequency of FPGA device is 100 MHz.

Fig. 6.20 Circle and ring displayed on the screen

Solution 6.5 We will make the VHDL implementation using components. Three components will be used for the design. The components and their connections are shown in Fig. 6.21. First component is the clock generation unit. For SVGA, 40 MHz pixel clock is used. For clock generator, IP cores can be used. For the generation of the pixel clock, programmable IP cores named as digital clock manager can be used. All the vendors which produce FPGA development platforms offer digital clock manager structures, and desired clock frequencies can be easily configured using these structures.

First, we will write the main VHDL code, i.e., top module, then write the VHDL codes for components "SVGA signal generation", i.e., component u2, and "VGA Circle", i.e., component u3. Overall, there are four VHDL programs, one of them is the main program, and three of them are the component implementations.

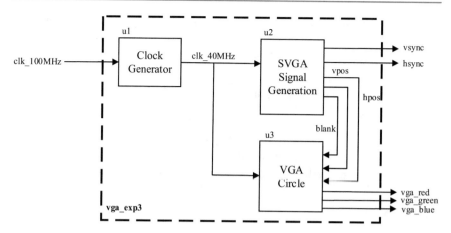

Fig. 6.21 Components of VHDL program for Example 6.5

In PR 6.41, the entity part of the main program is written, and signal objects used for pixel clock generation and display control operations are defined in the declarative part of the architecture unit.

```
library ieee;
use ieee.std_logic_1164.all;
use ieee.numeric_std.all;

entity vga_exp3 is
  port(clk_100MHz: in std_logic;
       Hsync, Vsync: out std_logic;
       vgaRed, vgaGreen, vgaBlue: out std_logic_vector(3 downto 0));
end vga_exp3;

architecture logic_flow of vga_exp3 is

  signal clk_40MHz: std_logic;
  signal blank: std_logic:='0';
  signal hpos, vpos: positive range 1 to 2048;
```

PR 6.41 Program 6.41

We need to define three components in the declarative part of the architecture. The first component used to generate 40 MHz pixel clock frequency is given in PR 6.42.

```
component clock_generator is
  port (clk_100MHz: in std_logic;
        clk_40MHz: out std_logic);
end component;
```

PR 6.42 Program 6.42

The second component is given in PR 6.43 where "blank" signal is used to activate or deactivate the display area. Referring to Table 6.1, maximum range for counter signals "hpos" and "vpos" is chosen as 2048.

```
component svga_signal_gen is
  port (clk: in  std_logic;
          blank: out std_logic;
          hsync, vsync: out std_logic;
          hpos, vpos: out positive range 1 to 2048);
  end component;
```

PR 6.43 Program 6.43

The third component, used for image generation, is written in PR 6.44.

```
component vga_circle is
  port (clk: in std_logic;
          blank_in: in std_logic;
          hpos, vpos: in positive range 1 to 2048;
          vga_red, vga_green, vga_blue: out std_logic_vector(3 downto 0);
          vga_blank: out std_logic);
  end component;
```

PR 6.44 Program 6.44

Port definitions are done considering the block diagram given in Fig. 6.20. Connections between components can be achieved using the **port map** function as in PR 6.45.

```
begin
  u1: clock_generator port map(
      clk_100MHz => clk_100MHz,
      clk_40MHz  => clk_40MHz);

  u2: svga_signal_gen port map(
      clk       => clk_40MHz,
      blank     => blank,
      hsync     => Hsync,
      vsync     => Vsync,
      hpos      => hpos,
      vpos      => vpos);

  u3: vga_circle port map(

      clk        => clk_40MHz,
      blank_in   => blank,
      hpos       => hpos,
      vpos       => vpos,
      vga_red    => vgaRed,
      vga_green  => vgaGreen,
      vga_blue   => vgaBlue);


  end logic_flow;
```

PR 6.45 Program 6.45

Combining all the program parts, we obtain the main VHDL program as in PR 6.46.

```vhdl
library ieee;
use ieee.std_logic_1164.all;
use ieee.numeric_std.all;

entity vga_exp3 is
 port(clk_100MHz: in std_logic;
      Hsync,Vsync: out std_logic;
      vgaRed, vgaGreen, vgaBlue: out
           std_logic_vector(3 downto 0));
end vga_exp3;

architecture logic_flow of vga_exp3 is

 signal clk_40MHz: std_logic;
 signal blank: std_logic:='0';
 signal hpos, vpos: positive range 1 to 2048;

 component clock_generator is
  port(clk_100MHz: in std_logic;
       clk_40MHz: out std_logic);
 end component;

 component svga_signal_gen is
  port(clk: in std_logic;
       blank: out std_logic;
       hsync, vsync: out std_logic;
       hpos,vpos: out positive range 1 to 2048);
 end component;

 component vga_circle is
  port(clk: in std_logic;
       blank_in: in std_logic;
       hpos,vpos: in positive range 1 to 2048;
       vga_red,vga_green,vga_blue: out
            std_logic_vector(3 downto 0);
       vga_blank: out std_logic);
 end component;

begin
 u1: clock_generator port map(
      clk_100MHz => clk_100MHz,
      clk_40MHz  => clk_40MHz);

 u2: svga_signal_gen port map(
      clk    => clk_40MHz,
      blank => blank,
      hsync => Hsync,
      vsync => Vsync,
      hpos  => hpos,
      vpos  => vpos);

 u3: vga_circle port map(
      clk         => clk_40MHz,
      blank_in    => blank,
      hpos        => hpos,
      vpos        => vpos,
      vga_red     => vgaRed,
      vga_green => vgaGreen,
      vga_blue   => vgaBlue);

end logic_flow;
```

PR 6.46 Program 6.46

After writing the main program as in PR 6.46, we can start writing the VHDL programs for the components used in the main program.

First, we will write the VHDL program for the component "svga_signal_gen", i.e., component u2. The monitor resolution is SVGA 800 × 600. The entity part and signal declarations in declarative part of the architecture are written as in PR 6.47. Referring to Table 6.1 the range limits for the counter parameters defined in the declarative part of the architecture are determined.

```
library ieee;
use ieee.std_logic_1164.all;
use ieee.numeric_std.all;

entity svga_signal_gen is
  port(clk: in std_logic;
       blank: out std_logic;
       hsync,vsync: out std_logic;
       hpos, vpos: out positive range 1 to 2048);
end svga_signal_gen;

architecture logic_flow of svga_signal_gen is
  signal x, y: integer range 0 to 2047:=0;
  signal act_pxl_hrzntl, act_pxl_vrtcl: positive range 1 to 2048:=1;
  signal hsync_sig: std_logic:='0';
  signal Hactive, Vactive: std_logic:='0';
begin
```

PR 6.47   Program 6.47

The VHDL program for generation of the horizontal synchronization signal is written in steps in PR 6.48, PR 6.49, PR 6.50, PR 6.51 and explained by small figures next to program parts.

```
horizontal_sync: process(clk)
begin
  if(rising_edge(clk)) then
    x<=x + 1;
    if (x<128) then
      hsync_sig<='0'; Hactive<='0';
```

PR 6.48   Program 6.48

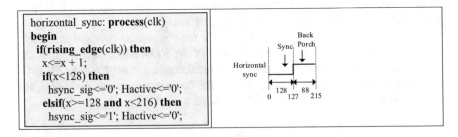

```
horizontal_sync: process(clk)
begin
  if(rising_edge(clk)) then
    x<=x + 1;
    if(x<128) then
      hsync_sig<='0'; Hactive<='0';
    elsif(x>=128 and x<216) then
      hsync_sig<='1'; Hactive<='0';
```

PR 6.49   Program 6.49

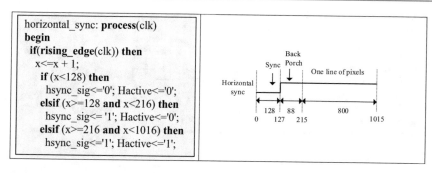

```
horizontal_sync: process(clk)
begin
  if(rising_edge(clk)) then
    x<=x + 1;
    if (x<128) then
      hsync_sig<='0'; Hactive<='0';
    elsif (x>=128 and x<216) then
      hsync_sig<= '1'; Hactive<='0';
    elsif (x>=216 and x<1016) then
      hsync_sig<='1'; Hactive<='1';
```

PR 6.50   Program 6.50

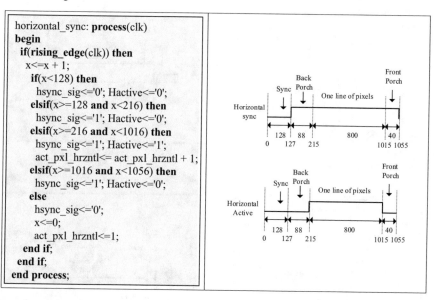

```
horizontal_sync: process(clk)
begin
  if(rising_edge(clk)) then
    x<=x + 1;
    if(x<128) then
      hsync_sig<='0'; Hactive<='0';
    elsif(x>=128 and x<216) then
      hsync_sig<='1'; Hactive<='0';
    elsif(x>=216 and x<1016) then
      hsync_sig<='1'; Hactive<='1';
      act_pxl_hrzntl<= act_pxl_hrzntl + 1;
    elsif(x>=1016 and x<1056) then
      hsync_sig<='1'; Hactive<='0';
    else
      hsync_sig<='0';
      x<=0;
      act_pxl_hrzntl<=1;
    end if;
  end if;
end process;
```

PR 6.51   Program 6.51

Generation of the vertical synchronization signal can be achieved in a similar manner. The sensitivity list of the process used for the generation of vertical synchronization signal contains horizontal synchronization signal, and at the rising edge of the horizontal synchronization signal vertical synchronization signal generation is made.

The VHDL program for generation of the vertical synchronization signal is written in steps in PR 6.52, PR 6.53, PR 6.54, PR 6.55 and explained by small figures next to program parts.

```
vertical_sync: process(hsync_sig)
begin
  if(rising_edge(hsync_sig)) then
    y<=y + 1;
    if(y<4) then
      Vsync<='0'; Vactive<='0';
```

PR 6.52   Program 6.52

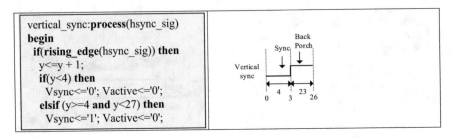

```
vertical_sync:process(hsync_sig)
begin
 if(rising_edge(hsync_sig)) then
  y<=y + 1;
  if(y<4) then
   Vsync<='0'; Vactive<='0';
  elsif (y>=4 and y<27) then
   Vsync<='1'; Vactive<='0';
```

PR 6.53 Program 6.53

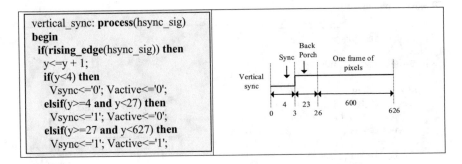

```
vertical_sync: process(hsync_sig)
begin
 if(rising_edge(hsync_sig)) then
  y<=y + 1;
  if(y<4) then
   Vsync<='0'; Vactive<='0';
  elsif(y>=4 and y<27) then
   Vsync<='1'; Vactive<='0';
  elsif(y>=27 and y<627) then
   Vsync<='1'; Vactive<='1';
```

PR 6.54 Program 6.54

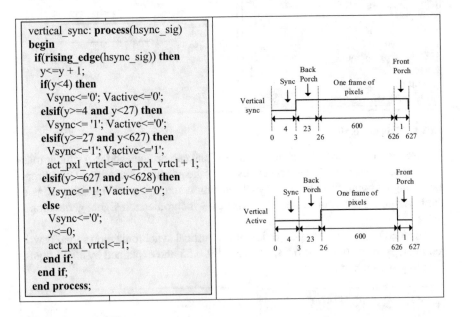

```
vertical_sync: process(hsync_sig)
begin
 if(rising_edge(hsync_sig)) then
  y<=y + 1;
  if(y<4) then
   Vsync<='0'; Vactive<='0';
  elsif(y>=4 and y<27) then
   Vsync<= '1'; Vactive<='0';
  elsif(y>=27 and y<627) then
   Vsync<='1'; Vactive<='1';
   act_pxl_vrtcl<=act_pxl_vrtcl + 1;
  elsif(y>=627 and y<628) then
   Vsync<='1'; Vactive<='0';
  else
   Vsync<='0';
   y<=0;
   act_pxl_vrtcl<=1;
  end if;
 end if;
end process;
```

PR 6.55 Program 6.55

Combining all the program units we get the overall VHDL program for the component "svga_signal_gen", i.e., component u2, as in PR 6.56.

```vhdl
library ieee;
use ieee.std_logic_1164.all;
use ieee.numeric_std.all;

entity svga_signal_gen is
  port(clk: in std_logic;
       blank: out std_logic;
       hsync: out std_logic;
       vsync: out std_logic;
       hpos, vpos: out positive range 1 to 2048);
end svga_signal_gen;

architecture logic_flow of svga_signal_gen is
  signal x, y: integer range 0 to 2047:=0;
  signal act_pxl_hrzntl, act_pxl_vrtcl: positive
                                range 1 to 2048:=1;
  signal hsync_sig: std_logic:='0';
  signal Hactive, Vactive: std_logic:='0';
begin
  hpos<=act_pxl_hrzntl;
  vpos<=act_pxl_vrtcl;
  Hsync<=hsync_sig;
  blank<=not(Hactive and Vactive);

  horizontal_sync:process(clk)
  begin
   if(rising_edge(clk)) then
    x<= x + 1;
    if(x<128) then
     hsync_sig<='0'; Hactive<='0';
    elsif(x>=128 and x<216) then
     hsync_sig<='1'; Hactive<='0';
    elsif(x>=216 and x<1016) then
     hsync_sig<='1'; Hactive<='1';
     act_pxl_hrzntl<=act_pxl_hrzntl + 1;
    elsif(x>=1016 and x<1056) then
     hsync_sig<='1'; Hactive<='0';
    else
     hsync_sig<='0';
     x<=0;
     act_pxl_hrzntl<=1;
    end if;
   end if;
  end process;

  vertical_sync: process (hsync_sig)
  begin
   if(rising_edge(hsync_sig)) then
    y<=y + 1;
    if(y<4) then
     Vsync<= '0'; Vactive<='0';
    elsif (y>=4 and y<27) then
     Vsync<='1'; Vactive<='0';
    elsif (y>=27 and y<627) then
     Vsync<='1'; Vactive<='1';
     act_pxl_vrtcl<= act_pxl_vrtcl + 1;
    elsif (y>=627 and y<628) then
     Vsync<='1'; Vactive<='0';
    else
     Vsync<='0';
     y<=0;
     act_pxl_vrtcl<=1;
    end if;
   end if;
  end process;
end logic_flow;
```

PR 6.56 Program 6.56

After completing the VHDL implementation of second component "svga_signal_gen", i.e., u2, we can start writing the VHDL implementation of third component "vga_circle", i.e., u3, which is used to draw the filled-circle and ring shapes. The entity part of the VHDL implementation for the component "vga_circle" is written in PR 6.57.

```
library ieee;
use ieee.std_logic_1164.all;
use ieee.numeric_std.all;

entity vga_circle is
  port(clk: in std_logic;
       blank_in: in std_logic;
       hpos,vpos: in positive range 1 to 2048;
       vga_red,vga_green,vga_blue: out std_logic_vector(3 downto 0));
end vga_circle;
```

PR 6.57   Program 6.57

In architecture body, we need a process to draw geometric shapes. The template for this process is given in PR 6.58.

```
architecture logic_flow of vga_circle is
begin
  circle_draw: process(clk)
  begin
    if(rising_edge(clk)) then
     if(blank_in='0') then
       --
       --Draw your shape  here
       --
      end if;
     end if;
   end process;
end logic_flow;
```

PR 6.58   Program 6.58

The shapes to be drawn by the process are depicted in Fig. 6.22 where the filled-circle can be expressed by the mathematical expression

$$(x-h)^2 + (y-k)^2 \leq r^2 \tag{6.1}$$

where h and k are the center coordinates of the circle and r is the radius of the circle. The radius of the filled-circle is chosen as 100, i.e., $r = 100$. Expanding the left side of (6.1), we obtain the mathematical expression for the filled-circle as

$$x^2 + y^2 + h^2 + k^2 - 2xh - 2yk \leq r^2 \tag{6.2}$$

which can be implemented using the port parameters in PR 6.57 as

$$\text{hpos}^2 + \text{vpos}^2 + 200^2 + 300^2 - 2 \times \text{hpos} \times 200 - 2 \times \text{vpos} \times 300 \leq 100 \times 100. \tag{6.3}$$

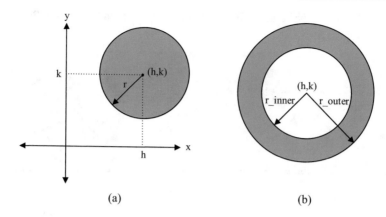

Fig. 6.22 Circle and ring shapes

Ring shape can be drawn using two circles with common center and with different radii as shown in Fig. 6.22b. In our example, the inner and outer radii of the circles for ring shape are taken as $r_{\text{inner}} = 100$ and $r_{\text{outer}} = 130$. The ring area can be mathematically described using

$$x^2 + y^2 + h^2 + k^2 - 2xh - 2yk \leq r_{\text{inner}}^2$$
$$x^2 + y^2 + h^2 + k^2 - 2xh - 2yk \geq r_{\text{outer}}^2$$

which can be implemented using the port parameters in PR 6.27 as

$$\text{hpos}^2 + \text{vpos}^2 + 200^2 + 300^2 - 2 \times \text{hpos} \times 200 - 2 \times \text{vpos} \times 300 \geq 100 \times 100$$
$$\text{hpos}^2 + \text{vpos}^2 + 200^2 + 300^2 - 2 \times \text{hpos} \times 200 - 2 \times \text{vpos} \times 300 \leq 130 \times 130$$

$$(6.4)$$

Using (6.2), we can write the VHDL code to draw the filled-circle as in PR 6.59.

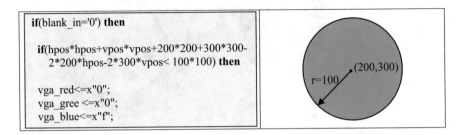

```
if(blank_in='0') then

    if(hpos*hpos+vpos*vpos+200*200+300*300-
      2*200*hpos-2*300*vpos< 100*100) then

    vga_red<=x"0";
    vga_gree <=x"0";
    vga_blue<=x"f";
```

PR 6.59   Program 6.59

Employing the formula (6.4), we can write the VHDL code to draw the rig shape as in PR 6.60.

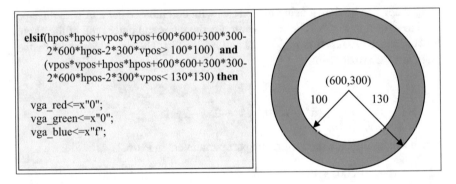

```
elsif(hpos*hpos+vpos*vpos+600*600+300*300-
      2*600*hpos-2*300*vpos> 100*100)  and
      (vpos*vpos+hpos*hpos+600*600+300*300-
      2*600*hpos-2*300*vpos< 130*130) then

    vga_red<=x"0";
    vga_green<=x"0";
    vga_blue<=x"f";
```

PR 6.60   Program 6.60

Combining all the program units, we obtain the overall VHDL implementation of the component "vga_circle", i.e., u3, as in PR 6.61.

```vhdl
library ieee;
use ieee.std_logic_1164.all;
use ieee.numeric_std.all;
entity vga_circle is
  port(clk: in std_logic;
       blank_in: in std_logic;
       hpos, vpos: in positive range 1 to 2048;
       vga_red, vga_green, vga_blue: out std_logic_vector(3 downto 0));
end vga_circle;

architecture logic_flow of vga_circle is
begin
  process(clk)
  begin
   if(rising_edge(clk)) then
    if(blank_in='0') then
     --Circle is created and filled by blue color
     if(hpos*hpos+vpos*vpos+200*200+300*300-2*200*hpos-2*300*vpos<=100*100) then
      vga_red<=x"0";
      vga_green<=x"0";
      vga_blue<=x"f";
     --Ring is created and filled by blue color
     elsif(hpos*hpos+vpos*vpos+600*600+300*300-2*600*hpos-2*300*vpos>=100*100)
          and(vpos*vpos+hpos*hpos+600*600+300*300-2*600*hpos-2*300*vpos<=130*130)
         then
      vga_red<=x"0";
      vga_green<=x"0";
      vga_blue<=x"f";
     else
      vga_red<=x"f";
      vga_green<=x"f";
      vga_blue<=x"f";
     end if;
    else
     vga_red<=x"0";
     vga_green<=x"0";
     vga_blue<=x"0";
    end if;
   end if;
  end process;
end logic_flow;
```

PR 6.61 Program 6.61

It is important to state that blanking intervals in terms of pixels should be defined by parameters in VHDL code. With this approach, VGA synchronization signals can be generated easily for all types of resolutions.

6.5.5 Generation and Display of Radar Screen

In this section, we will write a VHDL program for the generation and display of a typical radar screen. We will explain the subject through an example.

Example 6.6 Create a radar screen containing seven centered red circles with radii of 70, 100, 130, 160, 190, 220, and 250 pixels in length. The coordinates of the centers of the circles are (400, 300). Place a symbolic target, blue in color, on the radar screen to the coordinates (500, 200). Distance of the target from the center is 10 pixels in length. The background of the radar screen is white in color. Resolution of the monitor connected to FPGA board via a VGA cable is 800 × 600. In Fig. 6.23, the graphical illustration of the design is depicted.

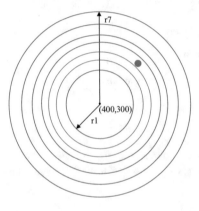

Fig. 6.23 A typical radar screen

Solution 6.6 The block diagram of the system to be designed is shown in Fig. 6.24 which is similar to the block diagram of the previous example. The only difference appears in the implementation of the third component u3. For this reason, we will only write the VHDL code for the third component u3, the other implementations for the other components can be used from the previous example.

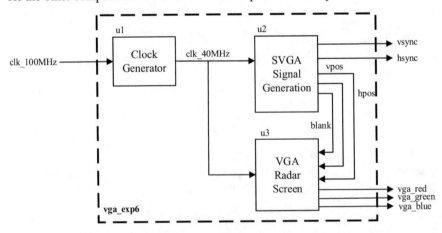

Fig. 6.24 Components of VHDL program for Example 6.6

The entity part for the implementation of third component "vga_radar_screen", i.e., component u3, is written in PR 6.62 where the same port definitions were made as in the previous example.

```
library ieee;
use ieee.std_logic_1164.all;
use ieee.numeric_std.all;

entity vga_radar_screen is
  port(clk: in std_logic;
       blank_in: in std_logic;
       hpos, vpos: in positive range 1 to 2048;
       vga_red, vga_green, vga_blue: out std_logic_vector(3 downto 0));
end vga_radar_screen;
```

PR 6.62 Program 6.62

The template of the process, which is used to draw the radar screen, is given in PR 6.63.

```
architecture logic_flow of vga_radar_screen is
begin
  radar_screen: process(clk)
  begin
    if(rising_edge(clk)) then
      if(blank_in='0') then
        --
        --Draw your shape  here
        --
      end if;
    end if;
  end process;
end logic_flow;
```

PR 6.63 Program 6.63

Radar screen contains seven circles, and the distance between the borders of two consecutive circles is 30 pixels in length. The VHDL code to draw the innermost circle is given in PR 6.64. The thickness of the circle border is 2 pixels in length.

```
--First ring r1=70 is drawn
if((vpos*vpos+hpos*hpos+400*400+300*300-
    2*400*(hpos)-2*300*vpos >70*70)  and
   (vpos*vpos+hpos*hpos+400*400+300*300-
    2*400*hpos-2*300*vpos <72*72)) then

vga_red<=x"f";
vga_green<=x"0";
vga_blue<=x"0";
```

PR 6.64 Program 6.64

The drawing of the second innermost circle is achieved in a similar manner in PR 6.65.

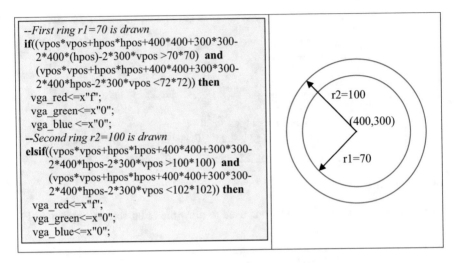

```
--First ring r1=70 is drawn
if((vpos*vpos+hpos*hpos+400*400+300*300-
    2*400*(hpos)-2*300*vpos >70*70) and
   (vpos*vpos+hpos*hpos+400*400+300*300-
    2*400*hpos-2*300*vpos <72*72)) then
 vga_red<=x"f";
 vga_green<=x"0";
 vga_blue <=x"0";
--Second ring r2=100 is drawn
elsif((vpos*vpos+hpos*hpos+400*400+300*300-
       2*400*hpos-2*300*vpos >100*100) and
      (vpos*vpos+hpos*hpos+400*400+300*300-
       2*400*hpos-2*300*vpos <102*102)) then
 vga_red<=x"f";
 vga_green<=x"0";
 vga_blue<=x"0";
```

PR 6.65   Program 6.65

The rest of the circles can be drawn on the radar screen in a similar manner. The drawing of the target symbol can be achieved using the VHDL code in PR 6.66 where the coordinates of the target is expressed using parameters, although numerical values are used for our example. The parameter "t_x" is the *x*-ordinate, and "t_y" is the *y*-ordinate for the center of the target.

```
--Target is drawn
elsif((vpos*vpos+hpos*hpos+t_x* t_x + t_y* t_y-
       2*t_x*hpos-2*t_y*vpos <10*10)) then
 vga_red<=x"0";
 vga_green<=x"0";
 vga_blue<=x"f";
```

PR 6.66   Program 6.66

Combining all the program segments, we get the VHDL implementation for the third component "vga_radar_screen" as in PR 6.67.

```vhdl
library ieee;
use ieee.std_logic_1164.all;
use ieee.numeric_std.all;
entity vga_radar_screen is
  port(clk,blank_in: in std_logic;
       hpos, vpos: in positive range 1 to 2048;
       vga_red, vga_green, vga_blue: out
                       std_logic_vector(3 downto 0));
end vga_radar_screen;

architecture logic_flow of vga_radar_screen is
signal t_x: positive range 1 to 2048:=500;
signal t_y: positive range 1 to 2048:=200;
begin
 process(clk)
 begin
  if(rising_edge(clk)) then
   if(blank_in='0') then
   --First ring r1=70 is drawn
    if((vpos*vpos+hpos*hpos+400*400+300*300-
        2*400*(hpos)-2*300*vpos>70*70) and
       (vpos*vpos+hpos*hpos+400*400+300*300-
        2*400*hpos-2*300*vpos<72*72)) then
     vga_red<=x"f";
     vga_green<=x"0";
     vga_blue<=x"0";
     --Second ring r2=100 is drawn
    elsif((vpos*vpos+hpos*hpos+400*400+300*300-
        2*400*hpos-2*300*vpos>100*100) and
       (vpos*vpos+hpos*hpos+400*400+300*300-
        2*400*hpos-2*300*vpos<102*102)) then
     vga_red<=x"f";
     vga_green<=x"0";
     vga_blue<=x"0";
     --Third ring r3=130 is drawn
    elsif((vpos*vpos+hpos*hpos+400*400+300*300-
        2*400*hpos-2*300*vpos>130*130) and
       (vpos*vpos+hpos*hpos+400*400+300*300-
        2*400* hpos-2*300*vpos<132*132)) then
     vga_red<=x"f";
     vga_green<=x"0";
     vga_blue<=x"0";
     --Fourth ring r4=160 is drawn
    elsif((vpos*vpos+hpos*hpos+400*400+300*300-
        2*400*hpos-2*300*vpos>160*160) and
       (vpos*(vpos)+ hpos*hpos+ 400*400+
300*300-
        2*400*hpos-2*300*vpos<162*162)) then

     vga_red<=x"f";
```

```vhdl
     vga_green<= x"0";
     vga_blue <= x"0";
     --Fifth ring r5=190 is drawn
    elsif((vpos*vpos+hpos*hpos+400*400+300*300-
        2*400*hpos-2*300*vpos>190*190) and
       (vpos*vpos+hpos*hpos+ 400*400 +300*300-
        2*400*hpos -2*300*vpos< 192*192)) then
     vga_red<=x"f";
     vga_green<=x"0";
     vga_blue<=x"0";
     --Sixth ring r6=220 is drawn
    elsif((vpos*vpos+hpos*hpos+400*400+300
        *300-2*400*hpos-2*300*vpos>220*220) and
       (vpos*vpos+hpos* hpos+400*400+ 300*300-
        2*400*hpos-2*300*vpos< 222*222)) then
     vga_red<=x"f";
     vga_green<=x"0";
     vga_blue<=x"0";
     --Last ring r7=250 is drawn
    elsif((vpos*vpos+hpos*hpos+400*400+300*
        300-2*400*hpos-2*300*vpos>250*250) and
       (vpos*vpos+hpos*hpos +400*400+ 300*300-
        2*400*hpos-2*300*vpos< 252*252)) then
     vga_red<=x"f";
     vga_green<=x"0";
     vga_blue<=x"0";
     --Target is drawn
    elsif((vpos*vpos+hpos*hpos+t_x*t_x+t_y*
        t_y-2*t_x* hpos-2*t_y*vpos<10*10)) then
     vga_red<=x"0";
     vga_green<=x"0";
     vga_blue<=x"f";
    else
     vga_red<=x"f";
     vga_green<=x"f";
     vga_blue<=x"f";
    end if;
   else
    vga_red<=x"0";
    vga_green<=x"0";
    vga_blue<=x"0";
   end if;
  end if;
 end process;
end logic_flow;
```

PR 6.67 Program 6.67

When the VHDL program in PR 6.67 is run, we get shapes similar to the one shown in Fig. 6.25.

Fig. 6.25 Output of Example 6.6

6.6   High-Definition Multimedia Interfacing (HDMI) and VHDL Implementation of HDMI

HDMI (High-definition Multimedia Interface) is introduced as an alternative to DVI and VGA interfaces for data transmission. In HDMI, different from DVI and VGA, audio data can be transmitted along with the video data

HDMI interface contains three types of communication channels which are display data channel (DDC), transition minimized differential signaling (TMDS) channel, and consumer electronics control (CEC) channel. Display data channel employs I²C bus specification, and it is used to get HDMI sink devices' properties from the HDMI source device.

TMDS is the main channel that carries video/audio data. In TMDS, 8-bit information data packets are encoded using 8b/10b encoding method as 10-bit packets which are used to carry digital data. 8b/10b encoding algorithm reduces the effects of electromagnetic interference. As the cable length increases, the transmitted data signal faces more degradation. TMDS encoding can be considered as a technique for data protection.

Three different data packet types are transmitted over TMDS lines, and these data packet types are video data packets, audio and auxiliary data packets, and control data packets. In this section, we will only consider the transmission of video data packets and control data packets over TMDS lines. The third channel type of HDMI is the consumer electronics control channel. This channel is used to control up to 15 compatible devices that are connected through HDMI lines to the same device. The pinout diagram for HDMI connector is depicted in Fig. 6.26.

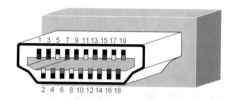

HDMI Pinout While Looking into Plug									
1 TMDS Data2+	**3** TMDS Data2-	**5** TMDS Data1 Shield	**7** TMDS Data0+	**9** TMDS Data0-	**11** TMDS Clock Shield	**13** CEC	**15** SCL	**17** DDC / CEC / HEC Ground	**19** Hot Plug Detect
	2 TMDS Data2 Shield	**4** TMDS Data1+	**6** TMDS Data1-	**8** TMDS Data0 Shield	**10** TMDS Clock+	**12** TMDS Clock-	**14** reserved	**16** SDA	**18** +5V power, 50 mA max

Fig. 6.26 HDMI pinout

6.6.1 TDMS Communication Channel

For VHDL implementation of HDMI, we need to know the details of TDMS channel. In this section, we will explain TMDS channel structure in detail. A TMDS channel, or communication block, consists of three parts, which are 8b/10b encoder, serializer, and clock manager, as depicted in Fig. 6.27.

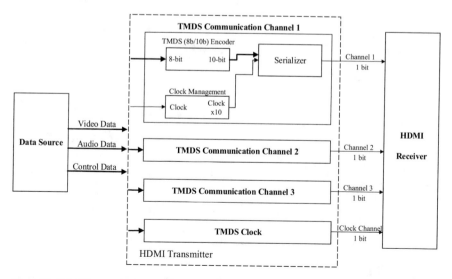

Fig. 6.27 HDMI transmitter

6.6.2   8b/10b Encoder

8b/10b encoding is introduced by IBM in 1983, and it is used in digital visual inter-face (DVI) specification developed by Digital Working Group. Using 8b/10b encod-ing it is possible to obtain DC balance on the transmission wires. DC balance is achieved keeping the number of ones and zeros equal. Due to DC balance, the noise margin is lowered. TMDS encoding algorithm is given in Fig. 6.28.

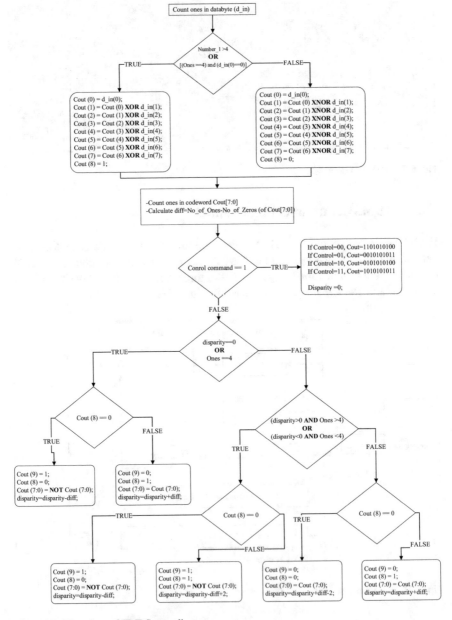

Fig. 6.28 Flowchart of TMDS encoding

Let us solve an example to illustrate the TMDS encoding operation.

Example 6.7   At the input of the TMDS encoder, we have the hexadecimal number $d_{in} = 0xF2$. Calculate 10-bit number at the output of the TMDS encoder for the three consecutive clock cycles. Use the flowchart given in Fig. 6.28. Starting disparity is zero.

Solution 6.7   First Clock Cycle

1. The binary equivalent of $d_{in} = 0xF2$ is "11110010". There are five ones in d_{in}. Disparity is zero. Cout (0) = d_{in} (0) = **0**;
2. Cout (0) = d_{in} (0) = **0**;
   Cout(1) = Cout(0) **xnor** d_{in}(1) = 0 ⊙ 1 = **0**;
   Cout(2) = Cout(1) **xnor** d_{in}(2) = 0 ⊙ 0 = **1**;
   Cout(3) = Cout(2) **xnor** d_{in}(3) = 1 ⊙ 0 = **0**;
   Cout(4) = Cout(3) **xnor** d_{in}(4) = 0 ⊙ 1 = **0**;
   Cout(5) = Cout(4) **xnor** d_{in}(5) = 0 ⊙ 1 = **0**;
   Cout(6) = Cout(5) **xnor** d_{in}(6) = 0 ⊙ 1 = **0**;
   Cout(7) = Cout(6) **xnor** d_{in}(7) = 0 ⊙ 1 = **0**;
   Cout(8) = 0.
3. Cout(8 : 0) = 0 0000 0100;
   The difference between number of ones in Cout and number of zeros in Cout is calculated as

$$Diff = 1 - 7 = -6.$$

4. Disparity is still zero and Cout(8) equals to 0
   Cout(9) = 1;
   Cout(8) = 0;
   Cout(7 : 0) = NOT Cout(7 : 0) = > 1111 1011;
   Thus, Cout equals = 10 1111 1011 = x"2FB" and disparity is calculated as

$$Disparity = Disparity - diff = 0 - (-6) = 6.$$

Second Clock Cycle
1. The binary equivalent of $d_{in} = 0xF2$ is "11110010". There are five ones in d_{in}. Disparity is 6.
2. Cout (0) = d_{in} (0) = **0**;
   Cout(1) = Cout(0) **xnor** d_{in}(1) = 0 ⊙ 1 = **0**;
   Cout(2) = Cout(1) **xnor** d_{in}(2) = 0 ⊙ 0 = **1**;
   Cout(3) = Cout(2) **xnor** d_{in}(3) = 1 ⊙ 0 = **0**;
   Cout(4) = Cout(3) **xnor** d_{in}(4) = 0 ⊙ 1 = **0**;
   Cout(5) = Cout(4) **xnor** d_{in}(5) = 0 ⊙ 1 = **0**;
   Cout(6) = Cout(5) **xnor** d_{in}(6) = 0 ⊙ 1 = **0**;
   Cout(7) = Cout(6) **xnor** d_{in}(7) = 0 ⊙ 1 = **0**;
   Cout(8) = 0.

3. $Cout(8 : 0) = 0\ 0000\ 0100$;
   The difference between number of ones in Cout and number of zeros in Cout is calculated as $Diff = 1 - 7 = -6$.
4. Disparity is 6 and number of ones in $Cout(7 : 0)$ is 1.
5. $Cout(8)$ equals to 0.
   $Cout(9) = 0$;
   $Cout(8) = 0$;
   $Cout(7 : 0) = Cout(7 : 0) = > 0000\ 0100$;
   Hence, Cout equals $= 00\ 0000\ 0100 = x"004"$ and disparity is calculated as

$$Disparity = Disparity + diff - 2 = 6 - 6 - 2 = -2.$$

Third Clock Cycle

1. The binary equivalent of $d_{in} = 0xF2$ is "11110010". There are five ones in d_{in}. Disparity is −2.
2. $Cout\ (0) = d_{in}\ (0) = \mathbf{0}$;
   $Cout(1) = Cout(0)\ \mathbf{xnor}\ d_{in}(1) = 0 \odot 1 = \mathbf{0}$;
   $Cout(2) = Cout(1)\ \mathbf{xnor}\ d_{in}(2) = 0 \odot 0 = \mathbf{1}$;
   $Cout(3) = Cout(2)\ \mathbf{xnor}\ d_{in}(3) = 1 \odot 0 = \mathbf{0}$;
   $Cout(4) = Cout(3)\ \mathbf{xnor}\ d_{in}(4) = 0 \odot 1 = \mathbf{0}$;
   $Cout(5) = Cout(4)\ \mathbf{xnor}\ d_{in}(5) = 0 \odot 1 = \mathbf{0}$;
   $Cout(6) = Cout(5)\ \mathbf{xnor}\ d_{in}(6) = 0 \odot 1 = \mathbf{0}$;
   $Cout(7) = Cout(6)\ \mathbf{xnor}\ d_{in}(7) = 0 \odot 1 = \mathbf{0}$;
   $Cout(8) = 0$.
3. $Cout(8 : 0) = 0\ 0000\ 0100$;
   The difference between number of ones in Cout and number of zeros in Cout is calculated as

$$Diff = 1 - 7 = -6.$$

4. Disparity is −2 and number of ones in $Cout(7 : 0)$ is 1.
5. $Cout(8)$ equals to 0;
   $Cout(9) = 1$;
   $Cout(8) = 0$;
   $Cout(7 : 0) = \mathbf{not}\ Cout(7 : 0) = > 1111\ 1011$;
   Hence, Cout equals to $10\ 1111\ 1011 = x"2FB"$, and the disparity is calculated as

$$Disparity = Disparity - diff = -2 - (-6) = 4.$$

Thus, outputs of the TMDS encoder are $x"2FB"$, $x"004"$, and $x"2FB"$ for three consecutive clock cycles. The number of ones and zeros in two consecutive sequences are equal to each other.

6.6.3   Implementation of TMDS Encoder in VHDL

In this section, we will explain VHDL implementation of TMDS encoder through an example.

Example 6.8 Implement the TMDS encoder algorithm illustrated in Fig. 6.28 in VHDL. The TMDS encoder has the inputs, "d_{in}", i.e., data input, "blank", "control_command", and "clk". At the output of the TMDS encoder, there is 10-bit wide word "Cout".

Solution 6.8 The black-box representation of the TMDS encoder, to be implemented in VHDL, is given in Fig. 6.29 where there are four input and one output ports.

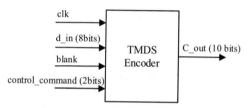

Fig. 6.29 Black-box representation of the TMDS encoder

Considering Fig. 6.29 and the algorithm depicted in Fig. 6.28, we write the entity part and declarative part of the architecture for VHDL implementation of TMDS encoder as in PR 6.68.

```
library ieee;
use ieee.std_logic_1164.all;
use ieee.numeric_std.all;

entity TDMS_encoder is
 port(clk: in std_logic;
       data: in std_logic_vector (7 downto 0);
       c: in std_logic_vector (1 downto 0);
       blank: in std_logic;
       encoded: out std_logic_vector (9 downto 0));
end TDMS_encoder;

architecture logic_flow of TDMS_encoder is
 signal ones, ones_count, zeros_count: integer range -8 to 8;
 signal disparity, diff: integer range -16 to 16:=0;
 signal Cout: std_logic_vector (9 downto 0);
 begin
```

PR 6.68 Program 6.68

In the body part of the architecture, the algorithm explained in Fig. 6.28 is implemented. The process to count the number of "1"s in the input byte is written in PR 6.69.

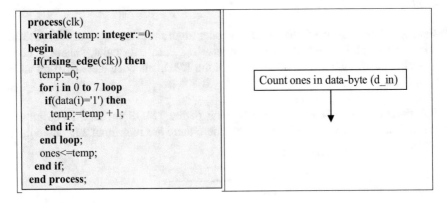

```
process(clk)
  variable temp: integer:=0;
begin
  if(rising_edge(clk)) then
    temp:=0;
    for i in 0 to 7 loop
      if(data(i)='1') then
        temp:=temp + 1;
      end if;
    end loop;
    ones<=temp;
  end if;
end process;
```

Count ones in data-byte (d_in)

PR 6.69   Program 6.69

In PR 6.70, calculation of TMDS encoder output is explained and implemented in VHDL.

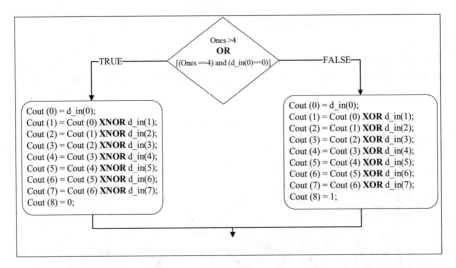

```
Cout(0)<=data(0);
Cout(1)<=data(1) xnor Cout(0) when (ones>4 or (ones=4 and data(0)='0')) else data(1) xor Cout(0);
Cout(2)<=data(2) xnor Cout(1) when (ones>4 or (ones=4 and data(0)='0')) else data(2) xor Cout(1);
Cout(3)<=data(3) xnor Cout(2) when (ones>4 or (ones=4 and data(0)='0')) else data(3) xor Cout(2);
Cout(4)<=data(4) xnor Cout(3) when (ones>4 or (ones=4 and data(0)='0')) else data(4) xor Cout(3);
Cout(5)<=data(5) xnor Cout(4) when (ones>4 or (ones=4 and data(0)='0')) else data(5) xor Cout(4);
Cout(6)<=data(6) xnor Cout(5) when (ones>4 or (ones=4 and data(0)='0')) else data(6) xor Cout(5);
Cout(7)<=data(7) xnor Cout(6) when (ones>4 or (ones=4 and data(0)='0')) else data(7) xor Cout(6);
Cout(8)<='0' when (ones>4 or (ones=4 and data(0)='0')) else '1';
```

PR 6.70   Program 6.70

The process unit used to count the difference between number of ones in Count and number of zeros in Count is explained and written in PR 6.71.

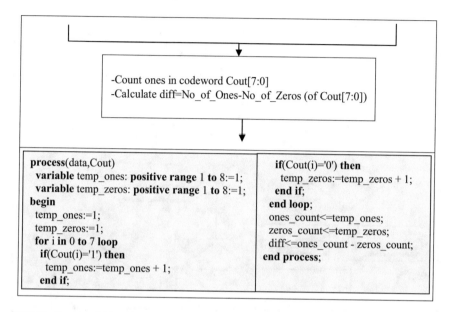

-Count ones in codeword Cout[7:0]
-Calculate diff=No_of_Ones-No_of_Zeros (of Cout[7:0])

```
process(data,Cout)                          if(Cout(i)='0') then
  variable temp_ones: positive range 1 to 8:=1;    temp_zeros:=temp_zeros + 1;
  variable temp_zeros: positive range 1 to 8:=1;   end if;
begin                                        end loop;
  temp_ones:=1;                              ones_count<=temp_ones;
  temp_zeros:=1;                             zeros_count<=temp_zeros;
  for i in 0 to 7 loop                       diff<=ones_count - zeros_count;
   if(Cout(i)='1') then                    end process;
    temp_ones:=temp_ones + 1;
   end if;
```

PR 6.71 Program 6.71

In PR 6.72, the conditional parts of the TMDS encoding algorithm is implemented in VHDL.

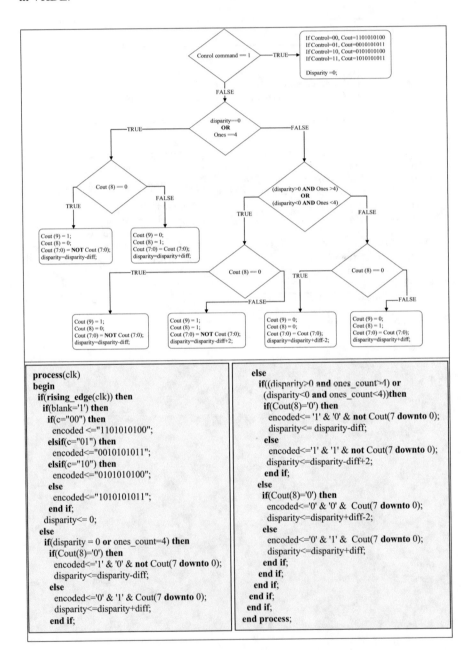

```
process(clk)
begin
  if(rising_edge(clk)) then
   if(blank='1') then
    if(c="00") then
      encoded <="1101010100";
    elsif(c="01") then
      encoded<="0010101011";
    elsif(c="10") then
      encoded<="0101010100";
    else
      encoded<="1010101011";
    end if;
    disparity<= 0;
   else
    if(disparity = 0 or ones_count=4) then
     if(Cout(8)='0') then
      encoded<='1' & '0' & not Cout(7 downto 0);
      disparity<=disparity-diff;
     else
      encoded<='0' & '1' & Cout(7 downto 0);
      disparity<=disparity+diff;
     end if;
```

```
   else
    if((disparity>0 and ones_count>4) or
       (disparity<0 and ones_count<4))then
     if(Cout(8)='0') then
      encoded<= '1' & '0' & not Cout(7 downto 0);
      disparity<= disparity-diff;
     else
      encoded<='1' & '1' & not Cout(7 downto 0);
      disparity<=disparity-diff+2;
     end if;
    else
     if(Cout(8)='0') then
      encoded<='0' & '0' & Cout(7 downto 0);
      disparity<=disparity+diff-2;
     else
      encoded<='0' & '1' & Cout(7 downto 0);
      disparity<=disparity+diff;
     end if;
    end if;
   end if;
  end if;
 end if;
end process;
```

PR 6.72 Program 6.72

Combining all the program units, we get the complete program for the VHDL implementation of TMDS encoding algorithm as in PR 6.73.

```vhdl
library ieee;
use ieee.std_logic_1164.all;
use ieee.numeric_std.all;
entity TDMS_encoder is
  port(clk: in  std_logic;
       data: in  std_logic_vector (7 downto 0);
       c: in  std_logic_vector (1 downto 0);
       blank: in  std_logic;
       encoded: out  std_logic_vector (9 downto 0));
end TDMS_encoder;
architecture logic_flow of TDMS_encoder is
  signal ones: integer range range -8 to 8;
  signal ones_count, zeros_count: integer range -8 to 8;
  signal disparity, diff: integer range -16 to 16:=0;
  signal Cout: std_logic_vector (9 downto 0;
begin
  process(clk)
    variable temp: integer:=0;
  begin
    if(rising_edge(clk)) then
      temp:=0;
      for i in 0 to 7 loop
        if(data(i) ='1') then
          temp:=temp + 1;
        end if;  end loop;
      ones<=temp;
    end if;end process;
  Cout(0)<=data(0);
  Cout(1)<=data(1) xnor Cout(0) when (ones>4 or
    (ones=4 and data(0)='0')) else data(1) xor Cout(0);
  Cout(2)<=data(2) xnor Cout(1) when (ones>4 or
    (ones=4 and data(0)='0')) else data(2) xor Cout(1);
  Cout(3)<=data(3) xnor Cout(2) when (ones>4 or
    (ones=4 and data(0)='0')) else data(3) xor Cout(2);
  Cout(4)<=data(4) xnor Cout(3) when (ones>4 or
    (ones=4 and data(0)='0')) else data(4) xor Cout(3);
  Cout(5)<=data(5) xnor Cout(4) when (ones>4 or
    (ones=4 and data(0)='0')) else data(5) xor Cout(4);
  Cout(6)<=data(6) xnor Cout(5) when (ones>4 or
    (ones=4 and data(0)='0')) else data(6) xor Cout(5);
  Cout(7)<=data(7) xnor Cout(6) when (ones>4 or
    (ones=4 and data(0)='0')) else data(7) xor Cout(6);
  Cout(8)<='0' when (ones>4 or (ones=4 and
    data(0)='0')) else '1';
  process(data, Cout)
    variable temp_ones: positive range 1 to 8:=1;
    variable temp_zeros: positive range 1 to 8:=1;
  begin
    temp_ones:=1;
    temp_zeros:=1;
    for i in 0 to 7 loop
      if(Cout(i)='1') then
        temp_ones:= temp_ones + 1;
      end if;
      if(Cout(i) = '0') then
        temp_zeros:= temp_zeros + 1;
      end if;
    end loop;
    ones_count<=temp_ones;
    zeros_count<=temp_zeros;
    diff<=ones_count - zeros_count;
  end process;
  process(clk)
  begin
    if(rising_edge(clk)) then
      if(blank='1') then
        if(c="00") then
          encoded<="1101010100";
        elsif (c="01") then
          encoded<="0010101011";
        elsif(c="10") then
          encoded<="0101010100";
        else
          encoded<="1010101011";
        end if;
        disparity<= 0;
      else
        if(disparity=0 or ones_count=4) then
          if(Cout(8)='0') then
            encoded<='1' & '0' & not Cout(7 downto 0);
            disparity<=disparity-diff;
          else
            encoded<='0' & '1' & Cout(7 downto 0);
            disparity<=disparity+diff;
          end if;
        else
          if((disparity>0 and ones_count>4) or
            (disparity<0 and ones_count<4))then
            if(Cout(8)='0') then
              encoded<= '1' & '0' & not Cout(7 downto 0);
              disparity<=disparity-diff;
            else
              encoded<= '1' & '1' & not Cout(7 downto 0);
              disparity<=disparity-diff+2;
            end if;
          else
            if(Cout(8)='0') then
              encoded<= '0' & '0' & Cout(7 downto 0);
              disparity<=disparity+diff-2;
            else
              encoded<='0' & '1' & Cout(7 downto 0);
              disparity<=disparity+diff;
          end if; end if; end if; end if;  end if;
      end process; end logic_flow;
```

PR 6.73 Program 6.73

6.6.4   Serializer in TMDS Communication Channel and Its VHDL Implementation

TMDS encoder takes 8-bit input data and produces 10-bit encoded data which appears at the output of the encoder on parallel lines. Using a parallel to serial converter, it is possible to serialize the parallel data. Serializer is a parallel to serial converter. In TDMS communication channel, serializer takes 10-bit parallel data and converts it to serial.

The clock frequency of the serializer should be ten times larger than the incoming data frequency. To achieve synchronization for serializer, two different strategies can be followed. In the first method, serializer clock frequency can be ×10 larger than the clock frequency of the encoder. In the second method, serializer can use ×5 larger clock frequency; however, in this case, both rising and falling edges of the larger clock are utilized for data transition. Both methods achieve serialization task. In Fig. 6.30, both methods are illustrated.

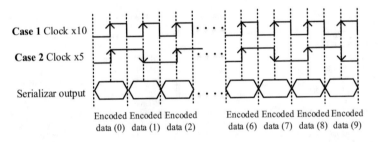

Fig. 6.30 Serializer clock types

We will consider the VHDL implementation of serializer used in TMDS communication channel through an example.

Example 6.9 Implement the serializer, which converts 10-bit parallel data to serial, in VHDL.

Solution 6.9 Data throughput speed of the serializer is ten times faster than input speed. The VHDL implementation of this serializer is given in PR 6.74.

```
library ieee;
use ieee.std_logic_1164.all;
entity serializer_10_to_1 is
  port(clk: in std_logic;
         clk_x10: in std_logic;
         data: in std_logic_vector (9 downto 0);
         serial_out: out std_logic);
end serializer_10_to_1;

architecture logic_flow of serializer_10_to_1 is
  signal new_data: std_logic_vector (9 downto 0);
begin
  process (clk_x10)
    variable count: integer range 0 to 10;
  begin
    if (clk_x10'event and clk_x10='1') then
      count:=count + 1;
      if (count=9) then
        new_data<=data;
      elsif (count=10) then
        count:=0;
      end if;
        serial_out<=new_data(count);
      end if;
  end process;
```

PR 6.74 Program 6.74

A simple PLL structure inside the FPGA can be used to obtain ×5 or ×10 faster clock signals to be used in TMDS communication channel.

6.6.5 VHDL Implementation of HDMI

Up to now, we considered parts of HDMI interface. Now, we can consider a complete VHDL program for data transmission via HDMI. It is seen in Fig. 6.27 that HDMI is used for data transmission, and HDMI gets audio, video, or control data from data sources separate from the HDMI. For this purpose, we can use our VGA source codes to generate image and control signals to be used for monitor display and send them via HDMI.

Let us explain the subject via an example.

Example 6.10 Design an HDMI transmitter in VHDL that is used to display a square shape of size 100 × 100 pixels in length on the monitor. The color of the square shape is red, and it should be displayed at the center of the monitor screen. Resolution of the monitor connected to FPGA board via an HDMI cable is 720P, i.e., 1280 × 720 pixels. 720P is the smallest of the high-definition resolutions supported by most of the monitors used today. In 720P, each frame contains 8Mbits.

Solution 6.10 Referring to Table 6.1, pixel configuration and clock management of the design can be performed. The block diagram of the design to be implemented in VHDL is depicted in Fig. 6.31.

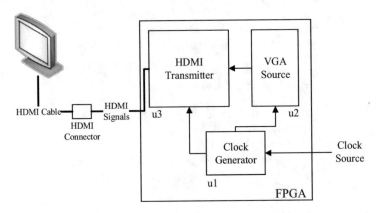

Fig. 6.31 Components of VHDL program for Example 6.10

In Fig. 6.31, display control and data signals are generated using VGA, and these signals are transmitted using TMDS communication channels of the HDMI. For the VHDL implementation of the FPGA side of Fig. 6.31, we will use components to write our VHDL codes. We will follow a top-down approach. First, we will write the main code involving component declarations and instantiations, then we will implement the components in VHDL.

The main program involving the components "HDMI transmitter", i.e., u3, "VGA source", i.e., u2, and "Clock generator", i.e., u1, is written in PR 6.75.

```
library ieee;
use ieee.std_logic_1164.all;
use ieee.numeric_std.all;

entity hdmi_display is
 port(Clk_100Mhz: in std_logic;
      hdmi_tx_rscl: out std_logic;
      hdmi_tx_rsda: inout std_logic;
      hdmi_tx_hpd: in std_logic;
      hdmi_tx_cec: inout std_logic;
      hdmi_tx_clk_p: out std_logic;
      hdmi_tx_clk_n: out std_logic;
      hdmi_tx_p: out std_logic_vector(2 downto 0);
      hdmi_tx_n: out std_logic_vector(2 downto 0));
end hdmi_display;
architecture logic_flow of hdmi_display is

 signal pixel_clk_x1: std_logic;
 signal pixel_clk_x10: std_logic;
 signal hpos, vpos: positive range 1 to 2048;
 signal vga_hsync: std_logic;
 signal vga_vsync: std_logic;
 signal vga_red: std_logic_vector(7 downto 0);
 signal vga_green: std_logic_vector(7 downto 0);
 signal vga_blue: std_logic_vector(7 downto 0);
 signal vga_blank: std_logic;

 component clock_generator is
 port(clk_100MHz: in std_logic;
      clk_75MHz: out std_logic;
      clk_750MHz: out std_logic);
 end component;

 component vga_source is
 port(pixel_clk: in std_logic;
      vga_hsync: out std_logic;
      vga_vsync: out std_logic;
      vga_red: out std_logic_vector(7 downto 0);
      vga_green: out std_logic_vector(7 downto 0);
      vga_blue: out std_logic_vector(7 downto 0);
      vga_blank: out std_logic);
 end component;
 component vga_to_hdmi is
 port(pixel_clk: in std_logic;
      pixel_clk_x10: in std_logic;
      vga_hsync: in std_logic;
      vga_vsync: in std_logic;
```

```
      vga_red: in std_logic_vector(7 downto 0);
      vga_green: in std_logic_vector(7 downto 0);
      vga_blue: in std_logic_vector(7 downto 0);
      vga_blank: in std_logic;
      hdmi_tx_rscl: out std_logic;
      hdmi_tx_rsda: inout std_logic;
      hdmi_tx_hpd: in std_logic;
      hdmi_tx_cec: inout std_logic;
      hdmi_tx_clk_p: out std_logic;
      hdmi_tx_clk_n: out std_logic;
      hdmi_tx_p: out std_logic_vector(2 downto 0);
      hdmi_tx_n: out std_logic_vector(2 downto 0));
 end component;
begin

u1: clock_generator port map (
      clk_100MHz => Clk_100Mhz,
      clk_75MHz  => pixel_clk_x1,
      clk_750MHz => pixel_clk_x10);

u2: vga_source port map (
      pixel_clk  => pixel_clk_x1,
      vga_hsync => vga_hsync,
      vga_vsync => vga_vsync,
      vga_red    => vga_red,
      vga_green => vga_green,
      vga_blue  => vga_blue,
      vga_blank => vga_blank);

u3: vga_to_hdmi port map (
      pixel_clk      => pixel_clk_x1,
      pixel_clk_x10  => pixel_clk_x10,
      vga_hsync     => vga_hsync,
      vga_vsync     => vga_vsync,
      vga_red       => vga_red,
      vga_green     => vga_green,
      vga_blue      => vga_blue,
      vga_blank     => vga_blank,
      hdmi_tx_rscl  => hdmi_tx_rscl,
      hdmi_tx_rsda  => hdmi_tx_rsda,
      hdmi_tx_hpd   => hdmi_tx_hpd,
      hdmi_tx_cec   => hdmi_tx_cec,
      hdmi_tx_clk_p => hdmi_tx_clk_p,
      hdmi_tx_clk_n => hdmi_tx_clk_n,
      hdmi_tx_p     => hdmi_tx_p,
      hdmi_tx_n     => hdmi_tx_n);
end logic_flow;
```

PR 6.75 Program 6.75

After writing the main program as in PR 6.75, we can start writing the VHDL programs for the components used in the main program.

The block diagram of the component "vga_source", i.e., u2, is depicted in Fig. 6.32.

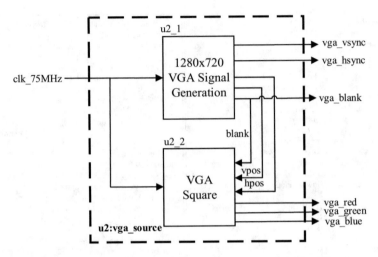

Fig. 6.32 Subcomponents for component u2

The component u2 explained in Fig. 6.32 can be implemented using subcomponents. The implementation of u2 can be achieved using a main program and two VHDL programs for the subcomponents u2_1 and u2_2. The main program is written in PR 6.76.

```
library ieee;
use ieee.std_logic_1164.all;

entity vga_source is
  port(pixel_clk: in std_logic;
       vga_hsync: out std_logic;
       vga_vsync: out std_logic;
       vga_red: out std_logic_vector(7 downto 0);
       vga_green: out std_logic_vector(7 downto 0);
       vga_blue: out std_logic_vector(7 downto 0);
       vga_blank: out std_logic);
end vga_source;

architecture logic_flow of vga_source is
  signal blank: std_logic:= '0';
  signal hpos, vpos: positive range 1 to 2048;

  component vga_gen_720p is
    port(clk: in std_logic;
         blank: out std_logic;
         hsync: out std_logic;
         vsync: out std_logic;
         hpos, vpos: out positive range 1 to 2048);
  end component;
```

```
component vga_square is
  port(clk: in std_logic;
       blank_in: in std_logic;
       hpos, vpos: in positive range 1 to 2048;
       vga_red: out std_logic_vector(7 downto 0);
       vga_green: out std_logic_vector(7 downto 0);
       vga_blue: out std_logic_vector(7 downto 0));
end component;
begin
  u2_1: vga_gen_720p port map (
       clk      => pixel_clk,
       blank    => blank,
       hsync    => vga_hsync,
       vsync    => vga_vsync,
       hpos     => hpos,
       vpos     => vpos);

  vga_blank<= blank;
  u2_2: vga_square port map (
       clk       => pixel_clk,
       blank_in  => blank,
       hpos      => hpos,
       vpos      => vpos,
       vga_red   => vga_red,
       vga_green => vga_green,
       vga_blue  => vga_blue);
end logic_flow;
```

PR 6.76 Program 6.76

Implementation of the sub-component "vga_gen_720", i.e., u2_1, is made in PR 6.77.

720P resolution contains 1650×750 pixels in total when front porch, back porch, and sync width parameters are considered, and 1280×720 of the pixels are used in active display screen area.

```vhdl
library ieee;
use ieee.std_logic_1164.all;
use ieee.numeric_std.all;

entity vga_gen_720p is
  port(clk: in std_logic;
       blank: out std_logic ;
       hsync, vsync: out std_logic ;
       hpos, vpos: out positive range 1 to 2048);
end vga_gen_720p;

architecture logic_flow of vga_gen_720p is

  signal x, y: integer range 0 to 2047:=0;
  signal act_pxl_hrzntl,act_pxl_vrtcl: positive range
  1 to 2048:=1;
  signal hsync_sig: std_logic:='0';
  signal Hactive, Vactive: std_logic:='0';

begin
  hpos<=act_pxl_hrzntl;
  vpos<=act_pxl_vrtcl;
  Hsync<=hsync_sig;
  blank<=not(Hactive and Vactive);
  p1: process (clk)
  begin
    if(rising_edge(clk)) then
      x<=x + 1;
      if (x<40) then
        hsync_sig<='0'; Hactive<='0';
      elsif (x>=40 and x<260) then
        hsync_sig<='1'; Hactive<='0';
      elsif (x>=260 and x<1540) then
        hsync_sig<='1'; Hactive<='1';
        act_pxl_hrzntl<=act_pxl_hrzntl + 1;
      elsif (x>=1540 and x<1650) then
        hsync_sig<='1'; Hactive<='0';
      else
        hsync_sig<='0';
        x<=0;
        act_pxl_hrzntl<=1;
      end if;
    end if;
  end process;

  p2: process (hsync_sig)
  begin
    if(rising_edge(hsync_sig)) then
      y<=y + 1;
      if (y<5) then
        Vsync<='0'; Vactive<='0';
      elsif (y>=5 and y<25) then
        Vsync<='1'; Vactive<='0';
      elsif (y>=25 and y<745) then
        act_pxl_vrtcl<=act_pxl_vrtcl + 1;
        Vsync<='1'; Vactive<='1';
      elsif (y>=745 and y<750) then
        Vsync<='1'; Vactive<='0';
      else
        Vsync<='0';
        y<=0;
        act_pxl_vrtcl<=1;
      end if;
    end if;
  end process;
end logic_flow;
```

PR 6.77 Program 6.77

Implementation of the sub-component "vga_square", i.e., u2_2, is made in PR 6.78.

```vhdl
library ieee;
use ieee.std_logic_1164.all;
use ieee.numeric_std.all;

entity vga_square is
  port(clk: in std_logic;
       blank_in: in std_logic;
       hpos, vpos: in positive range 1 to 2048;
       vga_red: out std_logic_vector(7 downto 0);
       vga_green: out std_logic_vector(7 downto 0);
       vga_blue: out std_logic_vector(7 downto 0));
end vga_square;

architecture logic_flow of vga_square is

  signal size: positive range 1 to 2048:=100;
  signal obj_X_pos: positive range 1 to 2048:=640;
  signal obj_Y_pos: positive range 1 to 2048:=360;
begin
  square_draw: process(clk)
  begin

    if(rising_edge(clk)) then
      if(blank_in='0') then
        if(0<=hpos+size-obj_X_pos) and
          (obj_X_pos+size-hpos>= 0) and
          (0<=vpos+size-obj_Y_pos) and
          (obj_Y_pos+size-vpos>= 0) then
          vga_red<=x"ff";
          vga_green<=x"00";
          vga_blue<=x"00";
        else
          vga_red<=x"ff";
          vga_green<=x"ff";
          vga_blue<= x"ff";
        end if;
      else
        vga_red<=x"00";
        vga_green<=x"00";
        vga_blue<=x"00";
      end if;
    end if;
  end process;
end logic_flow;
```

PR 6.78   Program 6.78

Now, VHDL implementation of the component "vga_to_hdmi", i.e., u3, will be made. In Fig. 6.33 detailed view of the component u3 is depicted. It is seen from Fig. 6.33 that there are two different subcomponents in u3, and these subcomponents are TMDS encoder and serializer.

The subcomponent TMDS encoder is used three times in u3, and the subcomponent serializer is used four times in u3. TMDS encoders are used to encode red, green, and blue color information. TMDS encoders accept 8-bit color data and produce 10-bit encoded data.

Serializers are used to convert parallel data at the output of the encoders to serial data. We will use the serializer explained in Example 6.9. Three of the serializers are used to encode the color data, and the fourth one is used to send clock signal which is ×10 faster than pixel clock.

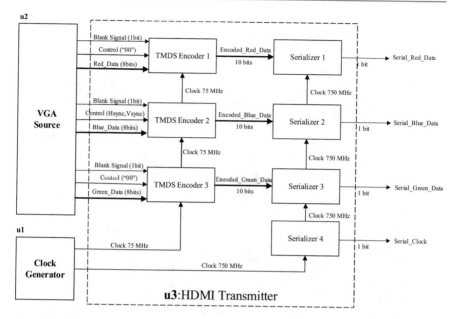

Fig. 6.33 Subcomponents for component u3

The VHDL program in PR 6.79 is written for the implementation of the component "vga_to_hdmi", i.e., u3.

HDMI uses differential signaling. Transition minimized differential signaling (TMDS) is used by HDMI. TMDS standard is used for video transmission. To achieve digital transmission through HDMI, buffer structures offered by FPGA vendors for their FPGA products to convert outputs of serializer into TMDS-based differential signals should be used. The name of buffer used for the FPGAs produced by XILINX company is OBUFDS. The use of OBUFDS is shown in PR 6.79. Besides, the pin format should be set to TMDS when pin assignments are made to test the code on FPGA board.

```vhdl
library ieee;
use ieee.std_logic_1164.all;
library UNISIM;
use UNISIM.Vcomponents.all;

entity vga_to_hdmi is
 port(pixel_clk: in std_logic;
        pixel_clk_x10: in std_logic;
        vga_hsync, vga_vsync: in std_logic;
        vga_red: in std_logic_vector(7 downto 0);
        vga_green: in std_logic_vector(7 downto 0);
        vga_blue: in std_logic_vector(7 downto 0);
        vga_blank: in std_logic;
        hdmi_tx_rscl: out std_logic;
        hdmi_tx_rsda: inout std_logic;
        hdmi_tx_hpd: in std_logic;
        hdmi_tx_cec: inout std_logic;
        hdmi_tx_clk_p: out std_logic;
        hdmi_tx_clk_n: out std_logic;
        hdmi_tx_p: out std_logic_vector(2 downto 0);
        hdmi_tx_n: out std_logic_vector(2 downto 0));
end vga_to_hdmi;

architecture logic_flow of vga_to_hdmi is

  component TDMS_encoder is
  port(clk: in std_logic;
        data: in std_logic_vector (7 downto 0);
        c: in std_logic_vector (1 downto 0);
        blank: in std_logic;
        encoded: out std_logic_vector (9 downto 0));
end component;

component serializer_10_to_1 is
  port(clk: in std_logic;
        clk_x10: in std_logic;
        data: in std_logic_vector (9 downto 0);
        serial_out: out std_logic);
end component;

  signal serial_clk: std_logic;
  signal serial_ch1: std_logic;
  signal serial_ch2: std_logic;
  signal serial_ch3: std_logic;
  signal c1_tmds_in: std_logic_vector(9 downto 0);
  signal c2_tmds_in: std_logic_vector(9 downto 0);
  signal c3_tmds_in: std_logic_vector(9 downto 0);
begin
  hdmi_tx_rsda<='Z';
  hdmi_tx_cec<='Z';
  hdmi_tx_rscl<='1';
  c1_tmds: TDMS_encoder port map(
                            clk    => pixel_clk,
                            data   => vga_blue,
                            c(1)   => vga_vsync,
                            c(0)   => vga_hsync,
                            blank  => vga_blank,
                            encoded => c1_tmds_in);

c2_tmds: TDMS_encoder port map (
            clk    => pixel_clk,
            data   => vga_green,
            c      => (others => '0'),
            blank  => vga_blank,
            encoded=> c2_tmds_in);

c3_tmds: TDMS_encoder port map (
            clk    => pixel_clk,
            data   => vga_red,
            c      => (others => '0'),
            blank  => vga_blank,
            encoded        => c3_tmds_in);

ser_ch1: serializer_10_to_1 port map (
            clk    => pixel_clk,
            clk_x10 => pixel_clk_x10,
            data    => c1_tmds_in,
            serial_out=> serial_ch1);

ser_ch2: serializer_10_to_1 port map (
            clk    => pixel_clk,
            clk_x10 => pixel_clk_x10,
            data    => c2_tmds_in,
            serial_out=> serial_ch2);

ser_ch3: serializer_10_to_1 port map (
            clk    => pixel_clk,
            clk_x10 => pixel_clk_x10,
            data    => c3_tmds_in,
            serial_out=> serial_ch3);

ser_clk: serializer_10_to_1 port map (
            clk    => pixel_clk,
            clk_x10 => pixel_clk_x10,
            data    => "0000011111",
            serial_out => serial_clk);

clk_buf: OBUFDS generic map ( IOSTANDARD => "TMDS_33", SLEW => "FAST")
  port map ( O => hdmi_tx_clk_p, OB => hdmi_tx_clk_n, I => serial_clk);

tx0_buf: OBUFDS generic map ( IOSTANDARD => "TMDS_33", SLEW => "FAST")
  port map ( O => hdmi_tx_p(0), OB => hdmi_tx_n(0), I => serial_ch1);

tx1_buf: OBUFDS generic map ( IOSTANDARD => "TMDS_33", SLEW => "FAST")
  port map ( O => hdmi_tx_p(1), OB => hdmi_tx_n(1), I => serial_ch2);

tx2_buf: OBUFDS generic map ( IOSTANDARD => "TMDS_33", SLEW => "FAST")
  port map ( O => hdmi_tx_p(2), OB => hdmi_tx_n(2), I => serial_ch3);

end logic_flow;
```

PR 6.79   Program 6.79

In the previous example, the image to be displayed on the monitor is formed using VHDL programming. However, it is difficult to generate complex images, such as the one shown in Fig. 6.34, via VHDL programming. For this reason, in the next example, we show how to generate image data in VHDL code from an image file.

Fig. 6.34 A nice image on monitor screen

Example 6.11 Save a grayscale image in block RAMs of the FPGA. Then, show this image on a screen with an HD720 resolution. Resolution of a pixel is represented with 8 bits.

Solution 6.11 The MATLAB code given in PR 6.80 converts a jpeg image to a grayscale image. The converted image consisting of 8-bit pixels is written into a text file.

```
clc; clear all; close all;
rgbImage=imread('your_image.jpg');
Gray_Image=rgb2gray(rgbImage);
fid=fopen('myTextFile.txt','wt');
size_gray=size(Gray_Image);
for i=1:size_gray(1)%%length of the rows
 for j=1:size_gray(2)%%length of the columns
  if(j==1)
    fprintf(fid,'(%d,',Gray_Image(i,j));
  elseif (j==size_gray(2) && i<size_gray(1))
    fprintf(fid,'%d),',Gray_Image(i,j));
  elseif (j==size_gray(2)&& i==size_gray(1))
    fprintf(fid,'%d )',Gray_Image(i,j));
  else
      fprintf(fid,'%d,',Gray_Image(i,j));
  end
 end
 if(i==size_gray(1))
    fprintf(fid,');');
 else
    fprintf(fid,'\n');
 end
end
fclose(fid);
imshow(Gray_Image);
```

PR 6.80 Program 6.80

Once you get the text file consisting of the integers, you can copy and paste comma-separated integers to the read part of the VHDL program shown in PR 6.81.

```vhdl
library ieee;
use ieee.std_logic_1164.all;
use ieee.numeric_std.all;

entity vga_picture is
  port(clk: in std_logic;
       blank_in: in std_logic;
       hpos, vpos: in positive range 1 to 2048;
       vga_red: out std_logic_vector(7 downto 0);
       vga_green: out std_logic_vector(7 downto 0);
       vga_blue: out std_logic_vector(7 downto 0));
end vga_picture;

architecture logic_flow of vga_picture is

  type ram_array is array(0 to 199, 0 to 199) of integer range 0 to 255;
  signal gray: ram_array;

begin
  gray <=((,........................,),
              content of 'myTextFile.txt'
              ........));;
  picture_show: process(clk)
  begin
   if(rising_edge(clk)) then
     if(blank_in='0') then
       if(540<=hpos and hpos<=739)  and (260<=vpos and vpos<=459) then
         vga_red<=std_logic_vector(to_unsigned(gray(vpos-260,hpos-540),8));
         vga_green<=std_logic_vector(to_unsigned (gray(vpos-260,hpos-540),8));
         vga_blue<=std_logic_vector(to_unsigned (gray(vpos-260,hpos-540),8));
       else
         vga_red<=x"ff";
         vga_green<=x"ff";
         vga_blue<=x"ff";
       end if;
     else
       vga_red<=x"00";
       vga_green<=x"00";
       vga_blue<=x"00";
     end if;
    end if;
  end process;
 end logic_flow;
```

PR 6.81   Program 6.81

Problems

1. Draw the horizontal synchronization signal of VGA format for resolution 640×480 and label the portions of the signal clearly.
2. Draw the vertical synchronization signal of VGA format for resolution 640×480 and label the portions of the signal clearly.
3. What is the period of pixel clock for 640×480 resolution.
4. What is the period of pixel frequency for 800×600 resolution.
5. Write VHDL, a program to display the digits 0, 1, 2, 3, 4, and 5 with 1 s delay on the monitor screen using VGA format.
6. Write VIIDL, a program to display the characters, A, B, C, and D with 1 s delay on the monitor screen. Assume that FPGA is connected to the monitor via HDMI.

Bibliography

1. Orhan Gazi, "A Tutorial Introduction to VHDL Programming," Springer, first edition, August 19, 2018, ISBN-13: 978-981132308.
2. Volnei A. Pedroni, "Circuit Design with VHDL," The MIT Press, third edition, April 14, 2020, ISBN-13: 978-0262042642.
3. Ricardo Jasinski, "Effective Coding with VHDL: Principles and Best Practice," The MIT Press, May 27, 2016, ISBN-13: 978-026203422.
4. Pong P. Chu, "FPGA Prototyping by Verilog Examples: Xilinx Spartan-3 Version," Wiley-Interscience, September 20, 2011, ISBN-13: 978-0470185322.
5. Volnei A. Pedroni, "Circuit Design and Simulation with VHDL," The MIT Press; second edition, September 17, 2010, ISBN-13: 978-026201433.
6. Andrew Rushton, "VHDL for Logic Synthesis," Wiley, third edition, April 25, 2011.
7. Eduardo Augusto Bezerra, Djones Vinicius Lettnin, "Synthesizable VHDL Design for FPGA," Springer, October 31, 2013, ISBN-13: 978-3319025469.
8. NXP Semiconductors, "UM10204, I2C-bus specification and user manual," rev. 6 - 4 April 2014.
9. Motorola, Inc., "SPI Block Guide," v03.06, original release date: 21 January 2000, revised 04 February 2003.

© The Author(s), under exclusive license to Springer Nature Switzerland AG 2021
O. Gazi, A. Ç. Arlı, *State Machines using VHDL*,
https://doi.org/10.1007/978-3-030-61698-4

Index

© The Author(s), under exclusive license to Springer Nature Switzerland AG 2021
O. Gazi, A. Ç. Arlı, *State Machines using VHDL*,
https://doi.org/10.1007/978-3-030-61698-4

Printed in the United States
by Baker & Taylor Publisher Services